McGraw-Hill's

500
Precalculus
Questions

Also in McGraw-Hill's 500 Questions Series

McGraw-Hill's 500 American Government Questions: Ace Your College Exams
McGraw-Hill's 500 College Algebra and Trigonometry Questions: Ace Your College Exams
McGraw-Hill's 500 College Biology Questions: Ace Your College Exams
McGraw-Hill's 500 College Calculus Questions: Ace Your College Exams
McGraw-Hill's 500 College Chemistry Questions: Ace Your College Exams
McGraw-Hill's 500 College Physics Questions: Ace Your College Exams
McGraw-Hill's 500 Differential Equations Questions: Ace Your College Exams
McGraw-Hill's 500 European History Questions: Ace Your College Exams
McGraw-Hill's 500 French Questions: Ace Your College Exams
McGraw-Hill's 500 Linear Algebra Questions: Ace Your College Exams
McGraw-Hill's 500 Macroeconomics Questions: Ace Your College Exams
McGraw-Hill's 500 Microeconomics Questions: Ace Your College Exams
McGraw-Hill's 500 Organic Chemistry Questions: Ace Your College Exams
McGraw-Hill's 500 Philosophy Questions: Ace Your College Exams
McGraw-Hill's 500 Physical Chemistry Questions: Ace Your College Exams
McGraw-Hill's 500 Psychology Questions: Ace Your College Exams
McGraw-Hill's 500 Spanish Questions: Ace Your College Exams
McGraw-Hill's 500 U.S. History Questions, Volume 1: Ace Your College Exams
McGraw-Hill's 500 U.S. History Questions, Volume 2: Ace Your College Exams
McGraw-Hill's 500 World History Questions, Volume 1: Ace Your College Exams
McGraw-Hill's 500 World History Questions, Volume 2: Ace Your College Exams
McGraw-Hill's 500 MCAT Biology Questions to Know by Test Day
McGraw-Hill's 500 MCAT General Chemistry Questions to Know by Test Day
McGraw-Hill's 500 MCAT Organic Chemistry Questions to Know by Test Day
McGraw-Hill's 500 MCAT Physics Questions to Know by Test Day

McGraw-Hill's

500

Precalculus
Questions

Ace Your College Exams

Sandra McCune
William Clark

New York   Chicago   San Francisco   Lisbon   London   Madrid   Mexico City
Milan   New Delhi   San Juan   Seoul   Singapore   Sydney   Toronto

3/11/13
ww
$16

The McGraw·Hill Companies

Copyright © 2013 by The McGraw-Hill Companies, Inc. All rights reserved. Printed in the United States of America. Except as permitted under the United States Copyright Act of 1976, no part of this publication may be reproduced or distributed in any form or by any means, or stored in a database or retrieval system, without the prior written permission of the publisher.

1 2 3 4 5 6 7 8 9 10 11 12 13 14 15 16   QFR/QFR   1 9 8 7 6 5 4 3 2

ISBN   978-0-07-178953-0
MHID   0-07-178953-7

e-ISBN   978-0-07-178954-7
e-MHID   0-07-178954-5

Library of Congress Control Number 2012933640

McGraw-Hill products are available at special quantity discounts to use as premiums and sales promotions or for use in corporate training programs. To contact a representative, please e-mail us at bulksales@mcgraw-hill.com.

This book is printed on acid-free paper.

CONTENTS

Introduction   ix

Chapter 1 Basic Algebraic Skills   1
Working with Real Numbers   1
  Questions 1–40
Complex Numbers   5
  Questions 41–55
The Cartesian Coordinate System   6
  Questions 56–65

Chapter 2 Precalculus Function Skills   9
Basic Function Concepts   9
  Questions 66–78
Operations with Functions   10
  Questions 79–120

Chapter 3 Graphs of Functions   13
Basic Graphing Concepts   13
  Questions 121–130
Increasing and Decreasing Behavior and Extrema   14
  Questions 131–137
Function Transformations   17
  Questions 138–165

**Chapter 4 Linear and Quadratic Functions,
Equations, and Inequalities   21**
Linear Functions   21
  Questions 166–174
Linear Equations and Inequalities   22
  Questions 175–182
Quadratic Functions   23
  Questions 183–196
Quadratic Equations and Inequalities   25
  Questions 197–207
Average Rate of Change and Difference Quotients for Linear and
Quadratic Functions   26
  Questions 208–220
Linear and Quadratic Function Models and Applications   27
  Questions 221–230

Chapter 5 Polynomial and Rational Functions 29

Polynomial Functions 29
　Questions 231–236
Remainder Theorem, Factor Theorem,
and Fundamental Theorem of Algebra 30
　Questions 237–245
Descartes' Rule of Signs and Rational Root Theorem 31
　Questions 246–250
Rational Functions 32
　Questions 251–256
Vertical, Horizontal, and Oblique Asymptotes
and Holes in Graphs of Rational Functions 33
　Questions 257–266
Behavior of Polynomial and Rational Functions Near $\pm\infty$ 36
　Questions 267–272

**Chapter 6 Exponential, Logarithmic,
and Other Common Functions 37**

Exponential Functions 37
　Questions 273–281
Logarithmic Functions 39
　Questions 282–294
Exponential and Logarithmic Equations 41
　Questions 295–302
Other Common Functions 42
　Questions 303–312

Chapter 7 Matrices and Systems of Linear Equations 45

Basic Concepts of Matrices 45
　Questions 313–322
Determinants and Inverses 46
　Questions 323–332
Solving Systems of Linear Equations Using Inverse Matrices 48
　Questions 333–334
Solving Systems of Linear Equations Using Cramer's Rule 49
　Questions 335–339
Solving Systems of Linear Equations Using Elementary
Row Transformations 50
　Questions 340–342

Chapter 8 Sequences, Series, and Mathematical Induction 53
Sequences 53
   Questions 343–362
Series 55
   Questions 363–372
Mathematical Induction 56
   Questions 373–382

Chapter 9 Trigonometric Functions 59
Angle Measurement 59
   Questions 383–392
Trigonometric Functions 60
   Questions 393–407
Trigonometric Graphs and Transformations 64
   Questions 408–432

Chapter 10 Analytic Trigonometry 69
Fundamental Trigonometric Identities 69
   Questions 433–442
Sum and Difference Identities 70
   Questions 443–455
Double- and Half-Angle Identities 71
   Questions 456–460
Trigonometric Equations 72
   Questions 461–472
Solving Triangles 74
   Questions 473–482

Chapter 11 Conic Sections 77
The Circle 77
   Questions 483–488
The Ellipse 78
   Questions 489–493
The Hyperbola 79
   Questions 494–497
The Parabola 80
   Questions 498–501

Answers 81

INTRODUCTION

Congratulations! You've taken a big step toward college success by purchasing *McGraw-Hill's 500 Precalculus Questions*. We are here to help you take the next step and prepare for your midterms, finals, and other exams so you can get the top grades you want!

This book gives you 500 questions that cover the most essential concepts in elementary and intermediate precalculus and in much of advanced precalculus. In the Answers section, for each problem, you'll find *a detailed way* to reach the solution. The questions and solutions will give you valuable independent practice to supplement your regular textbook and the ground you have already covered in your precalculus class.

This book and the others in this series were written by expert teachers who know the subject inside and out and can identify crucial information as well as the kinds of questions that are most likely to appear on your exams.

You might be the kind of student who spends weeks preparing for an exam. Or you might be the kind of student who puts off your exam preparation until the last minute. No matter what your preparation style, you will benefit from reviewing these 500 questions, covering the precalculus concepts you need to know to get top scores. These questions and solutions are the ideal preparation tool for any college precalculus test.

If you practice with all the questions and solutions in this book, we are certain you will build skills and gain the confidence needed to excel on your exams. Good luck!

—The Editors of McGraw-Hill Education

McGraw-Hill's

500

Precalculus

Questions

CHAPTER 1

Basic Algebraic Skills

Working with Real Numbers

1. List all the sets in the real number system to which the given number belongs.
   (A) 15
   (B) $\sqrt[3]{\dfrac{64}{125}}$
   (C) $-\pi$
   (D) -100
   (E) $\sqrt{3}$

2. Use interval notation to represent the indicated interval. State the interval type.
   (A) all numbers less than -4.5
   (B) $-12 \leq x \leq 28$
   (C)
   (D) all negative numbers
   (E) all real numbers

3. For each statement, tell which of the following properties of the real numbers, R, with respect to the operations of addition and multiplication is illustrated: Closure, Commutative, Associative, Additive Identity, Multiplicative Identity, Additive Inverse, Multiplicative Inverse, Distributive Property, Zero Factor Property.

(A) $0.45(2500) \in R$

(B) $\dfrac{3}{5} + \dfrac{1}{4} = \dfrac{1}{4} + \dfrac{3}{5}$

(C) $52 \cdot \dfrac{1}{52} = 1$

(D) $\left(1.5 + \dfrac{1}{8}\right) \in R$

(E) $23 + (11 + 9) = (23 + 11) + 9$

(F) $50(10 + 4) = 500 + 200$

(G) $-\sqrt{2} + \sqrt{2} = 0$

(H) $-99 \cdot 0 \cdot 100 = 0$

(I) $3\sqrt{5} + 7\sqrt{5} = (3 + 7)\sqrt{5}$

(J) $\left(8.5 \cdot \dfrac{5}{9}\right)\dfrac{9}{5} = 8.5\left(\dfrac{5}{9} \cdot \dfrac{9}{5}\right)$

4. Evaluate as indicated.

(A) $|20.9|$

(B) $\left|-7\dfrac{2}{3}\right|$

(C) $-|-80|$

(D) $\left|-\dfrac{\pi}{3}\right|$

(E) $|a|$, when a is negative

5. Insert $<$, $>$, or $=$ in the blank to make a true statement.

(A) $-|-3|$ _____ $-|2|$

(B) $-\left|-\dfrac{2}{5}\right|$ _____ $-\left|\dfrac{1}{2}\right|$

(C) $|-1.25|$ _____ $|1.25|$

(D) $-|-b|$ _____ $|b|, b \neq 0$

(E) $|-15|$ _____ $|-30|$

For questions 6 to 15, compute as indicated.

6. $-100 + -50$

7. $0.8 + -1.9$

8. $-34 - (-22)$

9. $\left(-\dfrac{35}{41}\right)\left(\dfrac{2}{5}\right)$

10. $\dfrac{24}{-3}$

11. $(-200)(-6)$

12. $-175.54 + 3.48 + 1.23$

13. $125 - (-437) - 80 + 359$

14. $(-0.25)(-600)(4.5)(-50)$

15. $\dfrac{-800}{-0.04}$

For questions 16 to 20, evaluate the given expression.

16. $(-3)^4$

17. -3^4

18. $49^{\frac{1}{2}}$

19. 5^{-3}

20. $32^{\frac{3}{5}}$

21. $(\sqrt{2})^0$

For questions 22 to 30, simplify the given expression. Assume that all variables are positive.

22. $x^5 x^3$

23. $\dfrac{y^7}{y^2}$

24. $(z^3)^4$

25. $\left(\dfrac{a}{b}\right)^4$

26. $\left(\dfrac{x}{y}\right)^{-2}$

27. $(a^3b^6)^{\frac{1}{3}}$

28. $(u^{-2})^4 v^7 v^{-9}$

29. $\dfrac{(x^2 y^{-5})^{-4}}{(x^5 y^{-2})^{-3}}$

30. $\dfrac{1}{2^{-1} + 3^{-1}}$

For questions 31 to 40, simplify using the order of operations. Refer to the following guidelines as needed.

Order of operations (PEMDAS):

1. Do computations inside <u>P</u>arentheses (or other grouping symbols).
2. Evaluate <u>E</u>xponential expressions (also, evaluate absolute value, square root, and other root expressions).
3. Perform <u>M</u>ultiplication and <u>D</u>ivision, in the order in which these operations occur from left to right.
4. Perform <u>A</u>ddition and <u>S</u>ubtraction, in the order in which these operations occur from left to right.

---

31. $(7+8)20 - 10$

32. $(-9^2)(5-8)$

33. $(20 - (-30))\left(-400^{\frac{1}{2}}\right)$

34. $8(-2) - \dfrac{15}{-5}$

35. $109 - \dfrac{20 + 22}{-6} - 4^3$

36. $(-2)^4 \cdot -3 - (15-4)^2$

37. $3(-11 + 3 \cdot 8 - 6 \cdot 3)^2$

38. $-15 - \dfrac{-10 - (3 \cdot -3 + 17)}{2}$

39. $\dfrac{6^2 - 8 \cdot 10 + 3^4 + 2}{3 \cdot 2 - 36 \div 12}$

40. $\dfrac{188 - 2(3^2 \cdot 2 - 3 \cdot 2^3)}{10^2}$

Complex Numbers

For questions 41 to 55, perform the indicated computation.

41. $(-8 - 5i) + (6 - 7i)$

42. $(9 + 4i) + (5 - 4i)$

43. $(1 - i\sqrt{5}) + (-7 + 2i\sqrt{5})$

44. $(16 - 5i) - (-3 + 7i)$

45. $\left(-\dfrac{1}{2} + \dfrac{3}{4}i\right) - \left(\dfrac{3}{2} + \dfrac{1}{4}i\right)$

46. $(4 + 3i)(10 - i)$

47. $(-2 - 5i)(4 - 8i)$

48. $(7 + 2i)(7 - 2i)$

49. $(\sqrt{3} - i\sqrt{5})(\sqrt{3} + i\sqrt{5})$

50. $(x + yi)\left(\dfrac{x}{x^2 + y^2} + \dfrac{-y}{x^2 + y^2}i\right)$

51. i^{302}

52. $\dfrac{1 - 2i}{3 + 4i}$

53. $\dfrac{4 - 2i}{2 + 3i}$

54. $(5 - 3i)^{-1}$

55. $(i)^{-1}$

The Cartesian Coordinate System

56. State the ordered pair of integers corresponding to each point in the coordinate plane shown.

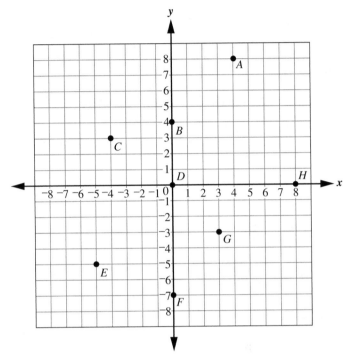

(A) _____

(B) _____

(C) _____

(D) _____

(E) _____

(F) _____

(G) _____

(H) _____

For questions 57 and 58, find the distance between each pair of points.

57. $(3, 4), (1, 7)$

58. $(-2, -3), (6, 0)$

For questions 59 and 60, find the midpoint between each pair of points.

59. $(7,10), (5,6)$

60. $(4,17), (-6,1)$

61. Circle the correct answer to make a true statement.

    (A) The change in y-coordinates between two points on a line is the (rise, run).

    (B) The change in x-coordinates between two points on a line is the (rise, run).

    (C) Lines that slant downward to the right have (negative, positive) slopes.

    (D) Lines that slant upward to the right have (negative, positive) slopes.

    (E) The slope of a horizontal line is (undefined, zero).

    (F) The slope of a vertical line is (undefined, zero).

For questions 62 to 65, find the slope of the line that contains the two given points.

62. $(-2,11), (4,-1)$

63. $(-3,-4), (0,0)$

64. $(-8,5), (-8,2)$

65. $(4,6), (-10,6)$

Precalculus Function Skills

Basic Function Concepts

66. Indicate whether the statement is true or false. Explain your reasoning.

    (A) $\{(-4,5),(-1,5),(0,5),(5,5)\}$ is a function.

    (B) $\{(x,y)\mid y^2 = 2x\}$ is a function.

    (C) If $(4,a)$ and $(4,b)$ are elements of a function, then $a = b$.

    (D) The domain of a function is a subset of that function.

    (E) In the function $f = \{(x,y)\mid y = 5x+3\}$, $y = f(x)$ is the image of x under f.

For questions 67 to 73, evaluate and simplify as indicated.

67. $y = f\left(\dfrac{3}{4}\right)$ when $f(x) = 28x - 10$

68. $y = f(-5)$ when $f(x) = x^2 + 1$

69. $y = f(-1)$ when $f(x) = 4x^5 + 2x^4 - 3x^3 - 5x^2 + x + 5$

70. $y = f\left(-\dfrac{\pi}{4}\right)$ when $f(x) = \dfrac{180\,|x|}{\pi}$

71. $y = f(3)$ when $f(x) = \dfrac{4x - 5}{x^2 + 1}$

72. $f(5x)$ when $f(x) = \dfrac{2x - 3}{x + 1}$

73. $f(x^2 + 1)$ when $f(x) = \dfrac{2x - 3}{x - 1}$

For questions 74 to 78, find the domain, D_f, and range, R_f, for the given real-valued functions. *Note:* In a real-valued function both the domain and range of the function consist of real numbers. All functions in this chapter are real-valued.

74. f defined by $f(x) = \sqrt{3x - 15}$

75. $f = \left\{ (5,12), (4,\sqrt{3}), \left(10, -\frac{3}{4}\right), (5.2, -1) \right\}$

76. g defined by $g(x) = \dfrac{5}{x + 4}$

77. h defined by $h(x) = |7x - 2|$

78. f defined by $f(x) = \sqrt{(x + 3)(x - 3)}$

Operations with Functions

For questions 79 to 82, using $f(x) = \dfrac{1}{x}$ and $g(x) = x^4$, write a simplified expression for the indicated function. Refer to the following definitions, as needed.

For all real numbers x such that $x \in D_f \cap D_g$:

$(f + g)(x) = f(x) + g(x)$;
$(f - g)(x) = f(x) - g(x)$;
$(fg)(x) = f(x) \cdot g(x)$; and

$\left(\dfrac{f}{g}\right)(x) = \dfrac{f(x)}{g(x)}$, where $g(x) \neq 0$.

79. $(f + g)(x)$
80. $(f - g)(x)$
81. $(fg)(x)$

82. $\left(\dfrac{f}{g}\right)(x)$

For questions 83 to 86, using $f(x) = x^2 + 3$ and $g(x) = \sqrt{x - 3}$, evaluate (if possible).

83. $(f + g)(4)$
84. $(f - g)(6)$
85. $(fg)(-1)$

86. $\left(\dfrac{f}{g}\right)(9)$

87. Let $f = \{(-5,1),(-4,-3),(-1,3),(2,5),(6,-4)\}$ and
$g = \{(-4,2),(-3,-3),(-1,3),(1,9),(3,-4)\}$; (a) find $f \circ g$ and (b) determine
$(f \circ g)(-4)$ (if possible).

88. Let $f = \{(-5,1),(-4,-3),(-1,3),(2,5),(6,-4)\}$ and
$g = \{(-4,2),(-3,-3),(-1,3),(1,9),(3,-4)\}$; (a) find $g \circ f$ and (b) determine
$(g \circ f)(6)$ (if possible).

89. Let $f = \{(-2,-8),(-1,3),(0,1),(1,4),(2,8)\}$ and
$g = \{(-2,-7),(-1,4),(0,-2),(1,2),(2,0),(3,8)\}$; (a) find $f \circ g$ and
(b) determine $(f \circ g)(1)$ (if possible).

90. Let $f = \{(-2,-8),(-1,3),(0,1),(1,4),(2,8)\}$ and
$g = \{(-2,-7),(-1,4),(0,-2),(1,2),(2,0),(3,8)\}$; (a) find $g \circ f$ and
(b) determine $(g \circ f)(2)$ (if possible).

For questions 91 to 95, use the given $f(x)$ and $g(x)$ to (a) write a simplified
expression for $(f \circ g)(x)$ and (b) determine the domain of $f \circ g$.

91. $f(x) = \sqrt{x+5}$ and $g(x) = 3x^2$

92. $f(x) = |x|$ and $g(x) = 9x - 4$

93. $f(x) = \sqrt{x}$ and $g(x) = 16x^2$

94. $f(x) = \dfrac{1}{x+1}$ and $g(x) = 5x$

95. $f(x) = \dfrac{1-x}{12}$ and $g(x) = 1 - 12x$

For questions 96 to 100, use $f(x) = \sqrt{x-4}$ and $g(x) = x^2 + 4$ to evaluate each
expression (if possible).

96. $(f \circ g)(9)$
97. $(f \circ g)(0)$
98. $(f \circ f)(40)$
99. $(f \circ f)(10)$
100. $(g \circ f)(5)$

For questions 101 to 105, determine whether the function is one-to-one.

101. $f = \{(1,4),(2,5),(3,6),(4,7),(5,8)\}$

102. $g = \left\{(-3,\pi),(0,0),(1,3.14),(2,4),\left(3,\dfrac{22}{7}\right)\right\}$

103. $h = \left\{\left(-1,\dfrac{1}{2}\right),\left(2,\dfrac{3}{4}\right),(3,4),(4,0.75),(5,1.5)\right\}$

104. $f = \{(x,y) \mid y = \sqrt{x}+1\}$

105. $g = \left\{(x,y) \mid y = \dfrac{5}{x^2}\right\}$

106. If $f = \{(0,1),(1,5),(2,9),(3,13)\}$, find f^{-1}.

For questions 107 to 111, find an equation that defines the inverse of the given function.

107. $y = f(x) = 3x+1$

108. $y = g(x) = (x+5)^3$

109. $h(x) = 7x$

110. $g(x) = \dfrac{x+2}{x-1}$

111. $y = f(x) = x^2$, where the domain of $f = \{x \mid x \geq 0\}$

For questions 112 to 115, evaluate $f^{-1}(x)$ at the value x for the function f defined as indicated.

112. $f = \{(0,1),(1,5),(2,9),(3,13)\}$, $x = 9$

113. $f(x) = \dfrac{5}{9}(x-32)$, $x = 0$

114. $f(x) = \dfrac{3}{4}x+5$, $x = 0.5$

115. $f(x) = \dfrac{x^3+6}{2}$, $x = -29$

For questions 116 to 120, determine whether the function f is even, odd, or neither.

116. $f(x) = x^3$

117. $g(x) = |x+3|$

118. $h(x) = 36x^2 - 10$

119. $g(x) = 2x^2 - x + 1$

120. $f(x) = 3x^4 - 10x^2 + 8$

CHAPTER 3

Graphs of Functions

Basic Graphing Concepts

In questions 121 to 124, use a graphing utility to graph the function.

121. Graph $y = x^2 - 4$ and state its domain and range.

122. Graph $y = \sqrt{x - 2}$ and state its domain and range.

123. Graph $y = x^3$ and state its domain and range.

124. Graph $y = \dfrac{2}{x^2 + 1}$ and state its domain and range.

For questions 125 and 126, determine whether the graph shown is the graph of a function. Explain your reasoning.

125.

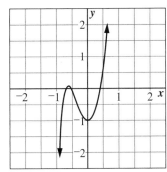

126.

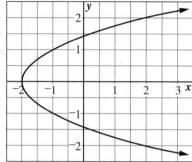

For questions 127 to 130, find the x- and y-intercepts.

127. $f(x) = 6x - 1$

128. $f(x) = 6x^2 + 5x - 4$

129. $p(x) = (x + 1)(x + 3)(x - 2)(x - 4)$

130. $g(x) = x^2 + 4$

Increasing and Decreasing Behavior and Extrema

For questions 131 to 133, use the graph of the function f to determine intervals where f is increasing, decreasing, or constant. Refer to the following definitions, as needed.

> A function f is strictly increasing on an interval I if, for every pair of numbers x_1 and x_2 in I, $f(x_1) < f(x_2)$ whenever $x_1 < x_2$; f is strictly decreasing on I if, for every pair of numbers x_1 and x_2 in I, $f(x_1) > f(x_2)$ whenever $x_1 < x_2$; f is constant on I if $f(x_1) = f(x_2)$ for every pair of numbers x_1 and x_2 in I.

131.

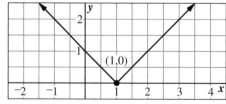

132.

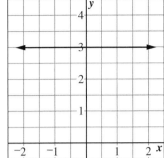

133.

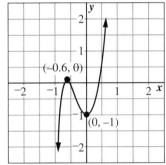

For questions 134 to 137, use the graph of the function f to determine any relative or absolute extrema. Refer to the following definitions, as needed.

$f(c)$ is an absolute minimum of f if $f(c) \leq f(x)$ for all x in D_f. Similarly, $f(c)$ is an absolute maximum of f if $f(c) \geq f(x)$ for all x in D_f. The minimum and maximum values are the extrema. $f(c)$ is a relative minimum of f if there exists an open interval containing c in which $f(c)$ is a minimum; similarly, $f(c)$ is a relative maximum of f if there exists an open interval containing c in which $f(c)$ is a maximum. If $f(c)$ is a relative minimum or maximum, it is called a relative extremum.

134.

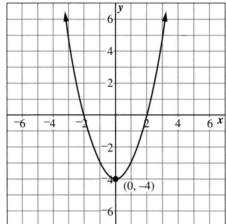

135.

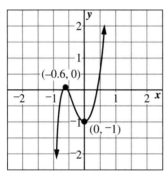

136.

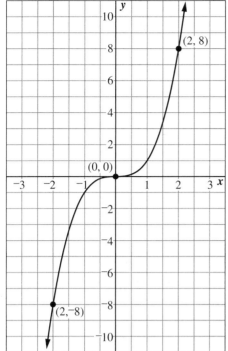

137.

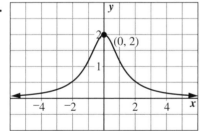

Function Transformations

For questions 138 to 142, write the equation for the graph of the function g that results when the given transformation is applied to the function f. Do not simplify the equation.

138. A vertical shift of 7 units up of the graph defined by $f(x) = \sqrt{x}$.

139. A horizontal shift of 7 units to the right of the graph defined by $f(x) = \sqrt{x}$.

140. A vertical shift of $\dfrac{2}{5}$ unit down of the graph defined by $f(x) = |x|$.

141. A horizontal shift of $\dfrac{2}{5}$ unit to the left of the graph defined by $f(x) = |x|$

142. A horizontal shift of 4 units to the right of the graph defined by $f(x) = \dfrac{1}{x}$

For questions 143 to 146, write an equation for a function g whose graph is congruent to the graph defined by $y = f(x)$ and that satisfies the given condition. Do not simplify the equation.

143. $f(x) = |x + 3|$, a reflection about the x-axis

144. $f(x) = |x + 3|$, a reflection about the y-axis

145. $f(x) = 5x^2 + 3x - 1$, a reflection about the x-axis

146. $f(x) = 5x^2 + 3x - 1$, a reflection about the y-axis

For questions 147 to 151, given that the graph of the function g is a dilation of the graph of the function f, describe the dilation as (a) a vertical stretch, (b) a vertical compression, (c) a horizontal stretch, or (d) a horizontal compression. Refer to the following definitions, as needed.

If $a > 1$, $y_2 = af(x)$ is a vertical stretch by a factor of a away from the x-axis of the graph $y_1 = f(x)$; whereas if $0 < a < 1$, $y_2 = af(x)$ is a vertical compression by a factor of a toward the x-axis of $y_1 = f(x)$.

If $b > 1$, $y_3 = f(bx)$ is a horizontal compression by a factor of $1/b$ toward the y-axis of $y_1 = f(x)$; whereas if $0 < b < 1$, $y_3 = f(bx)$ is a horizontal stretch by a factor of $1/b$ away from the y-axis of $y_1 = f(x)$.

---

147. $f(x) = \dfrac{1}{x^2}$, $g(x) = \dfrac{0.2}{x^2}$

148. $f(x) = \dfrac{1}{x^2}$, $g(x) = \dfrac{2}{x^2}$

149. $f(x) = x^2$, $g(x) = 15x^2$

150. $f(x) = \sqrt{x}$, $g(x) = \sqrt{0.2x}$

151. $f(x) = \sqrt{x}$, $g(x) = \sqrt{5x}$

For questions 152 to 155, write an equation for a function g whose graph is a dilation of the graph defined by $y = f(x)$, satisfying the condition given. Do not simplify the equation.

152. $f(x) = \dfrac{1}{x} - 3$, $g(x) = f(2x)$

153. $f(x) = \sqrt{x+1}$, $g(x) = 3f(x)$

154. $f(x) = x^3$, $g(x) = f\left(\dfrac{1}{2}x\right)$

155. $f(x) = \sqrt{x}$, $g(x) = 100 f(x)$

For questions 156 to 160, suppose that the graph of the function g is a dilation of the graph of the function f; if the given point is on the graph of f, give the coordinates of the corresponding point on the graph of g.

156. $f(x) = \dfrac{1}{x^2}$, $g(x) = 12\left(\dfrac{1}{x^2}\right)$, $\left(2, \dfrac{1}{4}\right)$

157. $f(x) = \dfrac{1}{x}$, $g(x) = \dfrac{2}{x}$, $\left(6, \dfrac{1}{6}\right)$

158. $f(x) = |x|$, $g(x) = |0.25x|$, $(-4, 4)$

159. $f(x) = x^2$, $g(x) = 10x^2$, $(-5, 25)$

160. $f(x) = \sqrt{x}$, $g(x) = \sqrt{3x}$, $(36, 6)$

For questions 161 to 165, write an equation for a function g whose graph is the result of transformations on the graph $y = f(x)$ that satisfy the given conditions. Do not simplify the equation.

161. $f(x) = \dfrac{1}{x^3}$; reflected about the y-axis, shifted right 5 units, up 3 units, and vertically stretched by a factor of 10

162. $f(x) = |x|$; reflected about the x-axis, shifted left 2.5 units, up 6.75 units, and vertically compressed by a factor of 0.25

163. $f(x) = x^3$; shifted right 7 units, down 8 units, vertically stretched by a factor of 5, and horizontally compressed by a factor of $\dfrac{1}{2}$

164. $f(x) = 2x^2 + 3x - 1$; shifted left 5 units and up 10 units

165. $f(x) = \sqrt{x}$; shifted left 3 units, down 5 units, horizontally stretched by a factor of 3, and vertically stretched by a factor of 6

CHAPTER 4

Linear and Quadratic Functions, Equations, and Inequalities

Linear Functions

For questions 166 to 170, (a) find the x-intercept(s) and y-intercept for the graph of f and (b) state the zero(s) of f. Refer to the following guidelines, as needed.

The graph of a linear function f defined by $f(x) = mx + b$ with $m \neq 0$ is a straight line with slope m that has exactly one x-intercept, $-\dfrac{b}{m}$, and exactly one y-intercept, b. To determine the zeros of a function, find all values x for which $f(x) = 0$. For $f(x) = mx + b$ with $m \neq 0$, the x-intercept, $-\dfrac{b}{m}$, is the only zero of f.

---

166. $f(x) = 3x + 5$

167. $f(x) = -\dfrac{3}{4}x - 7$

168. $f(x) = 10$

169. $f(x) = 0$

170. $f(x) = -2x - 5$

For questions 171 and 172, suppose f is a linear function whose graph contains the given point and has slope m. Write the equation $y = mx + b$ that defines f.

171. $m = \dfrac{3}{4}, (-8, -1)$

172. $m = -4, (-5, 2)$

For questions 173 and 174, suppose f is a linear function whose graph contains the given points. Write the equation $y = mx + b$ that defines f.

173. $(-2,11), (4,-1)$

174. $(4,6), (-10,6)$

Linear Equations and Inequalities

For questions 175 to 179, refer to the following guidelines, as needed.

Steps for Solving One-Variable Linear Equations

1. Remove grouping symbols, if any, and then simplify.
2. (optional) Eliminate fractions, if any, by multiplying both sides of the equation by the least common denominator of all the fractions in the equation; and then simplify.
3. Isolate the variable to one side of the equation, and then simplify.
4. If necessary, factor the side containing the variable so that one of the factors is the variable.
5. Divide both sides by the coefficient of the variable.

Note: You can combine steps and/or mentally complete steps; however, be careful when doing so because careless mistakes might occur.

---

175. Solve $x + 3(x - 2) = 2x - 4$ for x.

176. Solve $\dfrac{z}{5} - 3 = \dfrac{3}{10} - z$ for z.

177. Solve $\dfrac{x+3}{5} = \dfrac{x-1}{2}$ for x.

178. Solve $Ax + By = C$ for y. Write the answer in slope-intercept form.

179. Solve $x = \dfrac{y+2}{y-1}$ for y.

For questions 180 to 182, solve the linear inequality. Recall the following guideline as you work.

In solving linear inequalities proceed as you do with linear equations EXCEPT when you multiply or divide both sides by a negative number, reverse the direction of the inequality symbol.

---

180. $3x + 2 > 6x - 4$

181. $3x - 2 \leq 7 - 2x$

182. $\dfrac{x+3}{5} \geq \dfrac{x-1}{2}$

Quadratic Functions

In questions 183 to 185 for the function given, find (a) the y-intercept, and (b) the number of x-intercepts. Refer to the following guidelines, as needed.

The graph of $f(x) = ax^2 + bx + c$, $a \neq 0$, is a parabola (see Chapter 11 for a discussion of general parabolas) that has a y-intercept at c and x-intercepts at the real zeros (if any) of f. The zeros of f are the roots of the quadratic equation $ax^2 + bx + c = 0$. The discriminate $b^2 - 4ac$ of the quadratic equation gives you three possibilities for real zeros of f and thereby the x-intercepts: If $b^2 - 4ac > 0$, there are two real unequal zeros and, therefore, two x-intercepts; if $b^2 - 4ac = 0$, there is one real zero and, therefore, exactly one x-intercept; or if $b^2 - 4ac < 0$, there are no real zeros and, therefore, no x-intercepts.

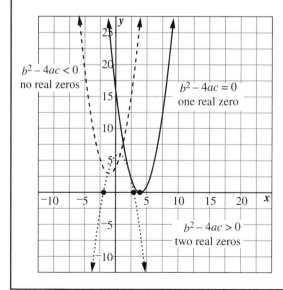

183. $f(x) = x^2 - 10x + 30$

184. $g(x) = 2x^2 + 2x - 5$

185. $h(x) = x^2 - 8x + 16$

In questions 186 to 188 for the function given, (a) state the vertex of its parabolic graph and (b) the equation of its axis of symmetry.

186. $f(x) = x^2 - 10x + 30$

187. $g(x) = 2x^2 + 2x - 5$

188. $h(x) = x^2 - 8x + 16$

For questions 189 to 191, (a) rewrite the equation for the quadratic function in vertex form, and (b) state the vertex of the parabolic graph.

189. $f(x) = 2x^2 - 12x + 17$

190. $h(x) = x^2 - 8x + 16$

191. $g(x) = -x^2 + 10x + 3$

192. Circle the correct words to make a true statement.

  (A) If $a > 0$, the parabola that is the graph $f(x) = ax^2 + bx + c$ opens (downward, upward); and if $a < 0$, the parabola opens (downward, upward).

  (B) When the leading coefficient of a quadratic function is negative, the y-coordinate of the vertex is the absolute (maximum, minimum) value of f.

  (C) When the leading coefficient of a quadratic function is positive, the y-coordinate of the vertex is the absolute (maximum, minimum) value of f.

In questions 193 to 196, for the graph of the given quadratic function, (a) state whether the parabola opens upward or downward, (b) find the vertex, (c) find the axis of symmetry, (d) find the extremum, (e) state the domain and range, and (f) state the intervals on which f is increasing or decreasing.

193. $f(x) = x^2 - 12x + 5$

194. $f(x) = x^2 + 4$

195. $g(x) = -3(x + 1)^2 - 7$

196. $g(x) = -x^2 + 10x + 3$

Quadratic Equations and Inequalities

When solving quadratics by factoring, refer to the following formulas, as needed.

$acx^2 + (ad + bc)x + bd = (ax + b)(cx + d)$	General trinomial
$x^2 + 2xy + y^2 = (x + y)^2$	Perfect square
$x^2 - y^2 = (x + y)(x - y)$	Difference of two squares
$x^2 + y^2 = (x + yi)(x - yi)$	Sum of two squares
	Note: Recall that $i^2 = -1$.

197. Solve $x^2 - x - 6 = 0$ by factoring.

198. Solve $x^2 + 6x - 1 = -5$ by completing the square.

199. Solve $3x^2 - 5x + 1 = 0$ by using the quadratic formula.

For questions 200 to 202, solve using any preferred method.

200. $x^2 - 3x + 2 = 0$

201. $9x^2 + 18x - 17 = 0$

202. $6x^2 - 12x + 7 = 0$

For questions 203 to 207, solve the quadratic inequality. Write the solution set in interval notation. Refer to the following guidelines, as needed.

To solve a quadratic inequality with $a > 0$, arrange terms so that only 0 is on the right side of the inequality and apply the following:

> If $ax^2 + bx + c = 0$ has two real roots, $ax^2 + bx + c$ is negative between them, positive to the left of the leftmost root, positive to the right of the rightmost root, and zero only at the roots.
> If $ax^2 + bx + c = 0$ has exactly one real root, $ax^2 + bx + c$ is zero at that root and positive elsewhere.
> If $ax^2 + bx + c = 0$ has no real roots, $ax^2 + bx + c$ is always positive.

Note: If you have a quadratic inequality in which $a < 0$, multiply both sides of the inequality by -1 and reverse the direction of the inequality to make $a > 0$.

203. $x^2 - x - 12 < 0$

204. $-x^2 + x + 12 < 0$

205. $x^2 - 10x + 30 > 0$

206. $x^2 - 10x + 25 \geq 0$

207. $x^2 - 5 \geq 0$

Average Rate of Change and Difference Quotients for Linear and Quadratic Functions

For questions 208 to 213, find the average rate of change of f on the given interval.

Note: The average rate of change of f as x goes from x_1 to x_2 is $\dfrac{f(x_2) - f(x_1)}{x_2 - x_1}$.

208. $f(x) = -2x - 5, [-3, 3]$

209. $f(x) = \dfrac{1}{2}x + 3, [-4, 8]$

210. $f(x) = 10, [-5, 5]$

211. $f(x) = kx, [-3, 5]$

212. $f(x) = x^2 + 8x - 5, [-8, -4]$

213. $f(x) = x^2 + 8x - 5, [-4, 8]$

214. Prove that the average rate of change of a linear function f defined by $f(x) = mx + b$ is constant over any interval $[c, d]$ and equals the slope m of the graph of f.

215. Prove that the average rate of change of a quadratic function f defined by $f(x) = ax^2 + bx + c$ over an interval $[m, n]$ equals $a(n + m) + b$.

For questions 216 to 219, find and simplify the difference quotient for the function.

Note: The difference quotient is the expression $\dfrac{f(x + h) - f(x)}{h}$, where $h \neq 0$. This expression is often used in calculus to find a general expression for the average rate of change of a function. It is the average rate of change of f as x goes from x to $x + h$.

216. $f(x) = 2x - 5$

217. $g(x) = 2x + 50$

218. $f(x) = x^2 + 3x$

219. $f(x) = ax^2 + bx + c$

220. Prove that the difference quotient of a linear function f defined by $f(x) = mx + b$ equals the slope m of the graph of f.

Linear and Quadratic Function Models and Applications

221. Suppose $f(t)$ is the distance traveled at time t by a car moving at a constant rate of speed of 65 miles per hour.

(A) Write a formula for $f(t)$.
(B) What is the distance traveled at time $t = 2\dfrac{3}{5}$ hours?

222. A tank contains 2,000 gallons of water. Suppose water is drained from the tank at a constant rate of 150 gallons per hour. Let $f(t)$ be the amount of water in the tank after t hours.

(A) Write a formula for $f(t)$.
(B) How much water is in the tank at $t = 3$ hours? *Hint:* The amount of water is *decreasing*, so the constant rate of change is negative.

223. A train is initially 245 miles from a city and moving toward the city at 60 miles per hour. Let $f(t)$ be the distance from the city at time t.

(A) Write a formula for $f(t)$.
(B) How far is the train from the city at $t = 2.5$ hours? *Hint:* The distance is *decreasing*, so the constant rate of change is negative.

224. Hooke's law states that $F = -kx$, where F is the force applied to a spring, x is the distance that the spring stretches from its natural state, and k is the constant of proportionality for the specific spring. Suppose for a certain spring the constant of proportionality is $24{,}000 \dfrac{\text{dynes}}{\text{cm}}$. *Note:* A dyne is a unit of force in the centimeter-gram-second (CGS) system. Let $F(x)$ be the force (in dynes) applied when the spring stretches a distance of x centimeters.

(A) Write a formula for $F(x)$.
(B) Find the force when the spring stretches 4 centimeters.

225. The monthly cost of driving a car includes both fixed and variable costs. Suppose that the fixed monthly cost for driving a certain car is $121 and the variable cost is 55.5 cents per mile driven. Let $f(x)$ be the monthly cost (in dollars) of driving this car for x miles.

(A) Write a formula for $f(x)$.
(B) Find $f(x)$ when $x = 180$ miles.

226. A farmer has 480 feet of fencing to enclose a rectangular pen for a donkey. Let $f(W)$ be the area of the rectangular pen expressed in terms of its width, W.

(A) Write a formula for $f(W)$.

(B) What dimensions for the rectangular pen will maximize the area for the donkey?

227. The height (in feet) above the ground for an object that is projected vertically upward with a velocity of 100 feet per second from an initial height of 30 feet is given by $h(t) = -16t^2 + 100t + 30$. Find the maximum height of the object above the ground.

228. Suppose the monthly revenue R in thousands of dollars that a company receives from producing x thousand boxes of designer watches is given by $R(x) = -4x^2 + 160x$. How many boxes of designer watches should the company produce each month to maximize $R(x)$?

229. Suppose a yearly cost function C in thousands of dollars is given by $C(x) = x^2 - 56x + 3000$. Find the minimum yearly cost.

230. Suppose an antique dealer determines that the value of a particular item depends on its age according to the formula $V(t) = -t^2 + 10t + 800$, where t is measured in hundreds of years. At what age does the item have maximum value?

CHAPTER 5

Polynomial and Rational Functions

Polynomial Functions

In questions 231 to 235, for the given polynomial function p, (a) determine the degree of $p(x)$, (b) describe the domain and range of the graph of p, (c) determine the zeros of p, (d) find the x-intercepts of the graph of p, and (e) find the y-intercept of the graph of p. Refer to the following guidelines, as needed.

A polynomial function p such that $p(x) = a_n x^n + a_{n-1} x^{n-1} + a_{n-2} x^{n-2} + \cdots + a_2 x^2 + a_1 x + a_0$, where n is a *nonnegative* integer and $a_n \neq 0$ is the leading coefficient, has degree n, the highest exponent of x. A constant polynomial's ($p(x) = c$, $c \neq 0$) degree is zero. The zero polynomial's ($p(x) = 0$) degree is undefined. The domain of p is R. If n is *odd*, the range is R; and if n is *even*, the range is a subset of R. The zeros of p are the roots of the equation $p(x) = 0$. That is, r is a zero of p if and only if $p(r) = 0$. If r is a real zero of p, then r is an x-intercept of the graph of p. The y-intercept of the graph of p is $p(0)$.

231. $p(x) = 3(x-1)(x+3)(x-4)(x+2)(x-2)$

232. $q(x) = (x^2+4)(x^2-5)(x^2-9)$

233. $g(x) = x^4 - 81$

234. $g(x) = -2x^2 - 9x + 5$

235. $f(x) = 3x + 5$

236. For the graph shown on the next page, identify (a) turning points and (b) relative or absolute extrema. Refer to the following guidelines, as needed.

The graph of a polynomial function will have a turning point (x, y) whenever the graph changes from increasing to decreasing or from decreasing to increasing. The y-value of a turning point is either a relative maximum or relative minimum value for the function. An nth-degree polynomial will have at most $n - 1$ turning points.

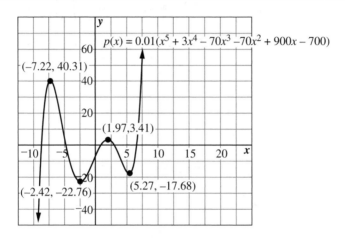

$$p(x) = 0.01(x^5 + 3x^4 - 70x^3 - 70x^2 + 900x - 700)$$

$(-7.22, 40.31)$

$(1.97, 3.41)$

$(-2.42, -22.76)$

$(5.27, -17.68)$

Remainder Theorem, Factor Theorem, and Fundamental Theorem of Algebra

237. Given $p(x) = 2x^3 - 5x^2 - 14x + 8$, use the remainder theorem to find $p(2)$.

238. Given $p(x) = 2x^3 - 5x^2 - 14x + 8$, use the remainder theorem to find $p(-2)$.

239. Use the factor theorem and the results in question 238 to factor $p(x) = 2x^3 - 5x^2 - 14x + 8$ completely.

240. The zeros of a polynomial function p of degree 4 and leading coefficient 5 are 3, −2, and $\pm\sqrt{2}$. Express $p(x)$ in factored form.

241. Suppose $g(x) = 2x^3 - 6x^2 - 2x + 6$ has zeros ±1 and 3. Express $g(x)$ in factored form.

242. Fill in the blank to make a true statement.

(A) The fundamental theorem of algebra states that, over the complex numbers, every polynomial equation of degree $n \geq 1$ has at least _____ root(s).

(B) If a root of a polynomial equation has multiplicity k, then that root will occur exactly _____ times in the list of all roots.

(C) The zeros of a polynomial p are the _____ of the equation $p(x) = 0$.

(D) The fundamental theorem of algebra guarantees that every polynomial of degree $n \geq 1$ has exactly _____ zeros, some of which might repeat.

(E) If $p(x)$ is a polynomial equation with real coefficients and $x + yi$ is a root of $p(x) = 0$, then its complex conjugate _____ is also a root of $p(x) = 0$.

For questions 243 and 244, (a) factor $p(x)$ completely, and (b) list all the zeros of p. Refer to the following formulas, as needed.

$acx^2 + (ad + bc)x + bd = (ax + b)(cx + d)$	General trinomial
$x^2 + 2xy + y^2 = (x + y)^2$	Perfect square
$x^2 - y^2 = (x + y)(x - y)$	Difference of two squares
$x^2 + y^2 = (x + yi)(x - yi)$	Sum of two squares
	Note: Recall that $i^2 = -1$.
$x^3 + y^3 = (x + y)(x^2 - xy + y^2)$	Sum of two cubes
$x^3 - y^3 = (x - y)(x^2 + xy + y^2)$	Difference of two cubes
$x = \dfrac{-b \pm \sqrt{b^2 - 4ac}}{2a}$	Quadratic formula

243. $p(x) = (x + 3)(x^2 - 5)(x^2 - 49)(x^2 + 49)$

244. $p(x) = (3x^2 - x - 10)(x^6 - 64)$

245. Find a polynomial equation $p(x)$ with real coefficients and with the least degree that has 3 and $2 - i$ as roots.

Descartes' Rule of Signs and Rational Root Theorem

For questions 246 to 248, discuss the positive or negative nature of the roots of the given polynomial equation. Refer to the following rule, as needed.

Descartes' rule of signs: If the polynomial equation $p(x) = 0$ has real coefficients, its maximum number of positive real roots either equals the number of sign variations in $p(x)$ or is less than that by an even number; and its maximum number of negative real roots either equals the number of sign variations in $p(-x)$ or is less than that by an even number. *Note:* When the terms of $p(x)$ are written in descending (or ascending) order of powers of x, a sign variation occurs if the coefficients of two consecutive terms have opposite signs (with missing terms being ignored).

246. $6x^4 + 7x^3 - 9x^2 - 7x + 3 = 0$

247. $x^3 - 1 = 0$

248. $x^5 + 4x^4 - 4x^3 - 16x^2 + 3x + 12 = 0$

249. List the possible rational roots for $2x^3 + x^2 - 13x + 6 = 0$.

250. Find the real roots for $2x^3 + x^2 - 13x + 6 = 0$.

Rational Functions

In questions 251 to 256, for the given rational function (a) state the domain, (b) find the zeros, (c) find the x-intercepts, and (d) find the y-intercepts. Refer to the following guidelines, as needed.

> A rational function f is defined as $f(x) = \dfrac{p(x)}{q(x)}$, where $p(x)$ and $q(x)$ are polynomials. Its domain is a subset of R, excluding all values of x for which $q(x) = 0$; and its range is a subset of R. When $f(x) = \dfrac{p(x)}{q(x)}$ is in simplified form, the zeros occur at x values for which $p(x) = 0$; the x-intercepts occur at real values for which $p(x) = 0$; and if 0 is in the domain, the y-intercept is $f(0)$.

251. $f(x) = \dfrac{3}{x}$

252. $g(x) = \dfrac{x^2 - 36}{3x^2 - x - 10}$

253. $f(x) = \dfrac{x^2 - 25}{x^2 - x - 6}$

254. $f(x) = \dfrac{x^3 - 8}{x - 4}$

255. $f(x) = \dfrac{x^3 - 8}{x^2 + 2x + 4}$

256. $h(x) = \dfrac{x + 1}{x^2 - 1}$

Vertical, Horizontal, and Oblique Asymptotes and Holes in Graphs of Rational Functions

For questions 257 to 261, (a) determine vertical asymptote(s) and (b) hole(s). Refer to the following definitions, as needed.

Suppose $f(x) = \dfrac{p(x)}{q(x)}$ is in simplified form, then

  a. The vertical line with equation $x = a$ is a vertical asymptote of the graph of f if $f(x)$ either increases or decreases without bound, as x approaches a from the left or right. To determine vertical asymptotes, set the denominator equal to 0 and solve. See the following graph.

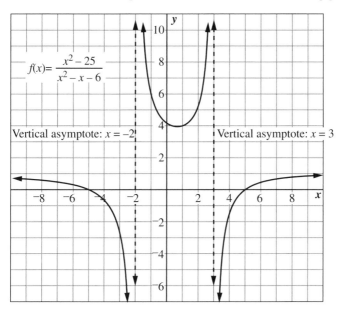

$$f(x) = \frac{x^2 - 25}{x^2 - x - 6}$$

Vertical asymptote: $x = -2$

Vertical asymptote: $x = 3$

(*continued*)

b. Upon inspection, if $p(x)$ and $q(x)$ have a common factor, $(x - h)$, that will divide out completely from the denominator when $f(x)$ is simplified, then the graph will have a "hole" at $(h, f(h))$, where $f(h)$ is calculated after $f(x)$ is simplified. See the following graph.

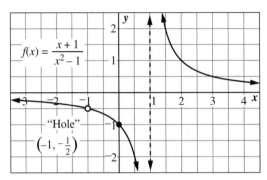

257. $f(x) = \dfrac{3}{x}$

258. $g(x) = \dfrac{x^2 - 4}{3x^2 - x - 10}$

259. $f(x) = \dfrac{x^3 - 8}{x - 4}$

260. $f(x) = \dfrac{x^3 - 8}{x^2 + 2x + 4}$

261. $h(x) = \dfrac{x + 3}{x^2 - 9}$

For questions 262 to 266, determine the horizontal or oblique asymptote. Refer to the following guidelines, as needed.

Suppose $f(x) = \dfrac{p(x)}{q(x)}$ is in simplified form, then a horizontal or oblique (slanted) line is an asymptote of the graph of f if $f(x)$ approaches the line as x approaches positive or negative infinity. See the following graph for an example of a horizontal asymptote.

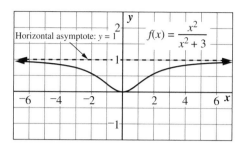

To determine horizontal or oblique asymptotes for $f(x) = \dfrac{p(x)}{q(x)}$, compare the degrees of the numerator and denominator polynomials. *Note:* The graph may have either a horizontal or an oblique asymptote—one or the other—but not both.

- If the numerator's degree is less than the denominator's degree, then the x-axis is a horizontal asymptote.
- If the numerator's degree equals the denominator's degree, then the graph has a horizontal asymptote at $y = \dfrac{a_n}{b_n}$, where a_n is the leading coefficient of $p(x)$ and b_n is the leading coefficient of $q(x)$.
- If the numerator's degree exceeds the denominator's degree by *exactly* 1, the graph has an oblique asymptote. To find the equation of the oblique asymptote, use division to rewrite $f(x) = \dfrac{p(x)}{q(x)}$ as quotient plus $\dfrac{\text{remainder}}{q(x)}$. The line with equation $y =$ quotient is the oblique asymptote.
- If the numerator's degree exceeds the denominator's degree by more than 1, the graph has neither a horizontal nor an oblique asymptote.

262. $f(x) = \dfrac{3}{x}$

263. $g(x) = \dfrac{x^2 - 4}{3x^2 - x - 10}$

264. $f(x) = \dfrac{x^3 - 8}{x - 4}$

265. $f(x) = \dfrac{x^2 + 3}{x - 1}$

266. $h(x) = \dfrac{x + 3}{x^2 - 9}$

Behavior of Polynomial and Rational Functions Near $\pm\infty$

In questions 267 to 272, for the given function f, (a) discuss the behavior of $f(x)$ as x approaches ∞, and (b) discuss the behavior of $f(x)$ as x approaches $-\infty$. Refer to the following guidelines, as needed.

As x approaches ∞ or $-\infty$,

a. $p(x) = a_n x^n + a_{n-1} x^{n-1} + a_{n-2} x^{n-2} + \cdots + a_2 x^2 + a_1 x + a_0$ behaves as does $a_n x^n$; and

b. $f(x) = \dfrac{p(x)}{q(x)}$, in simplified form, behaves as does $\dfrac{a_n x^n}{b_m x^m}$, where a_n is the leading coefficient of $p(x)$ and b_m is the leading coefficient of $q(x)$.

Note: Determining the behavior of functions as x nears $\pm\infty$ is a skill used in calculus.

---

267. $x^5 + 4x^4 - 4x^3 - 16x^2 + 3x + 12$

268. $f(x) = 6x^4 + 3x - 4$

269. $f(x) = -2x^4 + 100,000x^3$

270. $f(x) = \dfrac{3x - 4}{x^2 - 5x + 6}$

271. $f(x) = \dfrac{10x^2 - 9x - 6}{5x^2 + x - 12}$

272. $f(x) = \dfrac{2x^3 + 8x - 5}{-5x^2 + 6}$

CHAPTER 6

Exponential, Logarithmic, and Other Common Functions

Exponential Functions

For questions 273 to 275, find the function value for $f(x) = b^x$ ($b \neq 1, b > 0$). Refer to the following rules for exponents, as needed.

$x^1 = x$

$x^0 = 1, \ x \neq 0$

0^0 is undefined

$x^{-n} = \dfrac{1}{x^n}$

$\left(\dfrac{x}{y}\right)^{-1} = \dfrac{y}{x}$

$\left(\dfrac{x}{y}\right)^{-n} = \left(\dfrac{y}{x}\right)^n$

$x^{\frac{1}{n}} = \sqrt[n]{x}$

$x^{\frac{m}{n}} = (\sqrt[n]{x})^m = \sqrt[n]{x^m}$

$(x^n)^p = x^{np}$ (power of a power)

$\left(\dfrac{x}{y}\right)^p = \dfrac{x^p}{y^p}$ (power of a quotient)

$(xy)^p = x^p y^p$ (power of a product)

(continued)

$x^m x^n = x^{m+n}$ (product rule)

$\dfrac{x^m}{x^n} = x^{m-n}$ (quotient rule),

provided, in all cases, that *division by zero does not occur*, and when restricted to real numbers, that *even roots of negative quantities do not occur*.

273. $f(x) = 5^x$; $f(4)$

274. $f(x) = 27^x$; $f\left(\dfrac{4}{3}\right)$

275. $f(x) = \left(\dfrac{1}{2}\right)^x$; $f(-6)$

For questions 276 and 277, (a) state the domain and range, (b) find the zeros, (c) determine the asymptotes, (d) find the intercepts, (e) discuss increasing and decreasing behavior, and (f) discuss behavior as x approaches $\pm\infty$. Refer to the following guidelines, as needed.

The graph of $f(x) = b^x$ ($b \neq 1, b > 0$) is a smooth, continuous curve. The graph passes through the points $(0,1)$ and $(1,b)$ and is located in the first and second quadrants only. The domain is R, and the range is $(0,\infty)$. The y-intercept is 1. It has no x-intercepts. The x-axis is a horizontal asymptote.

Furthermore, the following hold:

- If $b > 1$, the function is increasing. As x approaches ∞, $f(x) = b^x$ approaches ∞. As x approaches $-\infty$, $f(x) = b^x$ approaches 0, but never reaches 0.
- If $0 < b < 1$, the function is decreasing. As x approaches ∞, $f(x) = b^x$ approaches 0, but never reaches 0. As x approaches $-\infty$, $f(x) = b^x$ approaches ∞.

276. $f(x) = 3^x$

277. $f(x) = \left(\dfrac{1}{4}\right)^x$

In questions 278 to 281, evaluate using the following properties of exponential functions, as needed.

For $f(x) = b^x$ ($b \neq 1, b > 0$),

$f(x) = b^x > 0$, for all real numbers

$f(0) = b^0 = 1$

$f(1) = b^1 = b$

$f(-x) = b^{-x} = \dfrac{1}{b^x}$

$f(u) \cdot f(v) = b^u \cdot b^v = b^{u+v} = f(u + v)$

$\dfrac{f(u)}{f(v)} = \dfrac{b^u}{b^v} = b^{u-v} = f(u - v)$

$(f(x))^p = (b^x)^p = b^{xp} = f(xp)$

One-to-one property: $f(u) = f(v)$ if and only if $u = v$; that is, $b^u = b^v$ if and only if $u = v$.

---

278. Suppose $f(x) = e^x$. Find $\dfrac{f(8)}{f(3)}$.

279. Suppose $f(x) = 2^x$. Find $f(3) \cdot f(2)$.

280. Suppose $f(x) = \left(\dfrac{7}{8}\right)^x$. Find $f(-1)$.

281. Suppose $f(x) = e^x$. If $e^u = e^{21}$, then $u = $ _____.

Logarithmic Functions

For questions 282 to 285, find the function value for $f(x) = \log_b x$, ($b \neq 1, b > 0$).

282. $f(x) = \log_5 x$; $f(625)$

283. $f(x) = \log_{27} x$; $f(81)$ *Hint:* Consider fractional exponents.

284. $f(x) = \log_{\frac{1}{2}} x$; $f(64)$

285. $h(x) = \log_{\frac{4}{9}} x$; $h\left(\dfrac{8}{27}\right)$

286. State the function g that defines the inverse of f.

(A) $f(x) = \log_6 x$

(B) $f(x) = (1.035)^x$

(C) $f(x) = \left(\dfrac{3}{4}\right)^x$

(D) $f(x) = \ln x$

(E) $f(x) = \log x$

For questions 287 and 288, (a) state the domain and range, (b) find the zeros, (c) determine the asymptotes, (d) find the intercepts, (e) discuss increasing and decreasing behavior, and (f) discuss behavior as x approaches 0 or ∞. Refer to the following guidelines, as needed.

> The graph of $f(x) = \log_b x$ ($b \neq 1, b > 0$) is a smooth, continuous curve. The graph passes through $(1,0)$ and $(b,1)$ and is located in the first and fourth quadrants only. The domain is $(0,\infty)$, and the range is R. The x-intercept is 1. It has no y-intercepts. The y-axis is a vertical asymptote.
>
> Furthermore, the following hold:
>
> • If $b > 1$, the function is increasing. As x approaches ∞, $f(x) = \log_b x$ approaches ∞. As x approaches 0, $f(x) = \log_b x$ approaches $-\infty$.
> • If $0 < b < 1$, the function is decreasing. As x approaches 0, $f(x) = b^x$ approaches ∞. As x approaches ∞, $f(x) = \log_b x$ approaches $-\infty$.

287. $f(x) = \log_6 x$

288. $g(x) = \log_{\frac{1}{2}} x$

In questions 289 to 292, evaluate using the following properties of logarithmic functions, as needed.

> For $f(x) = \log_b x$ ($b \neq 1, b > 0$),
>
> $f(1) = \log_b 1 = 0$
> $f(b) = \log_b b = 1$
> $f(b^x) = \log_b b^x = x$
> $f\left(\dfrac{1}{u}\right) = \log_b \dfrac{1}{u} = -\log_b u$
> $f(uv) = \log_b (uv) = \log_b u + \log_b v$

$$f\left(\frac{u}{v}\right) = \log_b\left(\frac{u}{v}\right) = \log_b u - \log_b v$$

$$f(u^p) = \log_b(u^p) = p \log_b u$$

One-to-one property: $f(u) = f(v)$ if and only if $u = v$; that is, $\log_b u = \log_b v$ if and only if $u = v$.

Change-of-base formula: $f(x) = \log_b x = \dfrac{\log_a x}{\log_a b} = \dfrac{\ln x}{\ln b} = \dfrac{\log_{10} x}{\log_{10} b}$ $(a \neq 1, a > 0)$.

289. Suppose $f(x) = \ln x$. Find $f(e^{10})$.

290. Suppose $g(x) = \log_2 x$. Find $g(64^{20})$.

291. Suppose $h(x) = \log x$. Find $h\left(\dfrac{100}{0.000001}\right)$.

292. Suppose $g(x) = \log_2 x$. Find $g(8 \cdot 32)$.

293. Suppose $f(x) = \ln x$. If $\ln u = \ln 450$, then $u =$ _____.

294. Use the property that $f(x) = \log_b x = \dfrac{\ln x}{\ln b}$ to express the given logarithm in terms of the natural logarithm function; then evaluate using the $\boxed{\text{LN}}$ key on your calculator (round to two decimal places, if needed).

(A) $\log_8(32,768)$
(B) $\log_{\frac{1}{5}}(0.0016)$
(C) $\log_2(4096)$
(D) $\log_{1.05}(2.5)$
(E) $\log_2(400)$

Exponential and Logarithmic Equations

For questions 295 to 299, solve, rounding to two decimal places when needed. Refer to the following inverse properties, as needed.

$$b^{\log_b x} = x \qquad \log_b b^x = x$$

$$e^{\ln x} = x \qquad \ln e^x = x$$

$$10^{\log_{10} x} = x \qquad \log 10^x = x$$

295. $8\log_2(3x-1)=256$

296. $\ln 8x = 3.5$

297. $4000(1.005)^x = 12,000$

298. $75e^{0.05x} = 150$

299. $e^{6x+1} = e^{4x-4}$

300. Radioactive elements undergo radioactive decay. The amount remaining, A, of a radioactive substance after t years is given by $A(t) = A_0\left(\dfrac{1}{2}\right)^{\frac{t}{k}}$, where A_0 is the initial amount of radioactive substance and k is its half-life. How much of a 20-gram sample of carbon-14, which has a half-life of 5730 years, remains after 5000 years?

301. The formula for the pH of an aqueous solution is given by $f(x) = -\log_{10} x$, where x is the hydronium ion concentration (in moles per liter) of the solution. Find the pH of diet soda Z, which has a hydronium concentration of 7.6×10^{-4}.

302. The compound interest formula is $P = P_0(1 + r)^t$, where r is the rate, compounded annually, and P is the value after t years of an initial investment of p_0. Suppose that the grandparents of a newborn child establish a trust fund account for the child with an investment of $50,000. Assuming no withdrawals and no additional deposits are made, approximately what interest rate compounded annually is needed to double the investment in 20 years?

Other Common Functions

303. Suppose $f(x) = \begin{cases} 2x+5 & \text{if } x < 0 \\ 3 & \text{if } x = 0 \\ \sqrt{2x+5} & \text{if } x > 0 \end{cases}$. Find each of the indicated function values.

(A) $f(-1)$

(B) $f\left(-\dfrac{5}{2}\right)$

(C) $f(0)$

(D) $f(5.5)$

(E) $f(3)$

304. Suppose $f(x) = [x]$, the greatest integer function. Find each of the following function values.

(A) $f(3.99)$
(B) $f(-3.99)$
(C) $f(0.005)$
(D) $f(-15)$
(E) $f(\pi)$

For questions 305 to 309, find the solution set. Refer to the following properties of the absolute value function, as needed.

$|x| \geq 0$

$|x| = |-x|$

$|xy| = |x||y|$

$\left|\dfrac{x}{y}\right| = \dfrac{|x|}{|y|}$

$|u + v| \leq |u| + |v|$

$\sqrt{x^2} = |x|$

$|x| = 0$ if and only if $x = 0$

If $c > 0$, then

$\quad |x| = c$ if and only if either $x = c$ or $x = -c$

$\quad |x| < c$ if and only if $-c < x < c$

$\quad |x| > c$ if and only if either $x < -c$ or $x > c$

Note: Properties involving $<$ and $>$ hold if you replace $<$ with $\leq$ and $>$ with $\geq$.

305. $\sqrt{x^2} = 9$

306. $|-2x - 3| < 15$

307. $|5x| \geq 40$

308. $-2|4x + 1| \geq -10$

309. $|x| < -1$

For questions 310 to 312, evaluate the power function f defined as $f(x) = x^a$, where a is a real number. Round to two places, when needed.

310. $f(x) = x^{-3}$; $f(10)$

311. $f(x) = x^{0.75}$; $f(256)$

312. $f(x) = x^{\sqrt{2}}$; $f(6)$

CHAPTER 7

Matrices and Systems of Linear Equations

Note: In this chapter, uppercase letters (A, B, C, ...) denote matrices, and lower-case letters (a, b, c, x, y, z, ...) denote scalars (numbers). Unless stated otherwise, all scalars are real numbers.

Basic Concepts of Matrices

313. What is the size of matrix $A = \begin{bmatrix} 2 & 2 \\ 0 & 1 \\ 2 & 5 \end{bmatrix}$?

314. In the matrix $A = [a_{ij}]_{3\times5} = \begin{bmatrix} 2 & -1 & 4 & 3 & 7 \\ 1 & 6 & -2 & 5 & 0 \\ -4 & 9 & 8 & -5 & -7 \end{bmatrix}$, identify the specified element.
  (A) a_{13}
  (B) a_{31}
  (C) a_{35}
  (D) a_{14}
  (E) a_{33}

315. Exhibit the indicated matrix.
  (A) The 3×1 column vector whose elements are 2, 3, and 8, in this order
  (B) The 1×4 row vector whose elements are 0, -2, 1, 5, in this order
  (C) The 3×3 matrix $A = [a_{ij}]_{3\times3}$ whose main diagonal elements are 3, -4, 5 and whose off-diagonal elements are 1's
  (D) $I_{3\times3}$
  (E) The 2×3 zero matrix

For questions 316 to 319, compute as indicated.

316. $\begin{bmatrix} 4 & 10 \\ -3 & 0 \end{bmatrix} + \begin{bmatrix} 3 & 2 \\ -2 & 3 \end{bmatrix}$

317. $\begin{bmatrix} 3 & 6 & 0 \\ 8 & -2 & -5 \\ 6 & -4 & 0 \end{bmatrix} - \begin{bmatrix} -1 & 6 & 4 \\ 0 & 3 & -1 \\ 5 & 0 & 3 \end{bmatrix}$

318. $2\begin{bmatrix} 3 & 1 & 1 \\ 1 & -4 & 1 \\ 1 & 1 & 5 \end{bmatrix}$

319. $\begin{bmatrix} 3 & 6 & 1 \end{bmatrix} \begin{bmatrix} 2 \\ -3 \\ 5 \end{bmatrix}$

320. $\begin{bmatrix} 3 & 6 & 1 \\ 0 & 5 & 4 \end{bmatrix} \begin{bmatrix} 2 & 4 \\ -3 & 1 \\ 5 & 0 \end{bmatrix}$

321. $\begin{bmatrix} 1 & 0 \\ 0 & 1 \end{bmatrix} \begin{bmatrix} 2 & 6 \\ -4 & 7 \end{bmatrix}$

322. Let $A = \begin{bmatrix} -4 & 5 \\ 1 & 2 \end{bmatrix}$ and $B = \begin{bmatrix} 3 & -1 \\ -6 & 2 \end{bmatrix}$.

    (A) Find AB.
    (B) Find BA.
    (C) Does $AB = BA$?

Determinants and Inverses

For questions 323 to 327, find the determinant of the indicated matrix.

323. $\begin{bmatrix} -4 & 5 \\ 1 & 2 \end{bmatrix}$

324. $\begin{bmatrix} 3 & -1 \\ -6 & 2 \end{bmatrix}$

325. $\begin{bmatrix} 3 & 4 & 2 \\ -2 & 8 & 2 \\ -2 & 1 & 5 \end{bmatrix}$

326. $\begin{bmatrix} 1 & 0 & 0 \\ 0 & 1 & 0 \\ 0 & 0 & 1 \end{bmatrix}$

327. $\begin{bmatrix} 1 & 4 \\ -3 & 1 \\ 6 & 0 \end{bmatrix}$

For questions 328 to 331, let $A = \begin{bmatrix} 4 & 1 \\ -3 & -2 \end{bmatrix}$ and $B = \begin{bmatrix} -1 & 1 \\ -2 & 4 \end{bmatrix}$.

328. Find (a) A^{-1} and (b) B^{-1}. Refer to the following guidelines, as needed.

The inverse of a square matrix A is a square matrix of the same size, A^{-1}, such that $AA^{-1} = A^{-1}A = I$. A simple method for finding the inverse of a 2×2 matrix is given by the following procedure:

In general, if $A = \begin{bmatrix} a & b \\ c & d \end{bmatrix}$, then $A^{-1} = \dfrac{1}{ad - bc}\begin{bmatrix} d & -b \\ -c & a \end{bmatrix} = \begin{bmatrix} \dfrac{d}{ad - bc} & \dfrac{-b}{ad - bc} \\ \dfrac{-c}{ad - bc} & \dfrac{a}{ad - bc} \end{bmatrix}$,

provided $ad - bc \neq 0$. In words, switch the elements on A's diagonal, negate the other two elements (but don't switch them), and then multiply the resulting matrix by $\dfrac{1}{|A|}$.

Caution: This method works for 2×2 matrices only. (The inverses for larger matrices are efficiently found using technological devices such as graphing calculators). Note that this procedure for finding the inverse of a 2×2 matrix A will not work if det $A = 0$. In fact, if the determinant of any square matrix is 0, the matrix will not have an inverse.

---

329. Find $(AB)^{-1}$.

330. Compute $A^{-1}B^{-1}$.

331. Compute $B^{-1}A^{-1}$.

332. Indicate whether the statement is true or false.

(A) $(AB)^{-1}$ equals $B^{-1}A^{-1}$.

(B) $(AB)^{-1}$ equals $A^{-1}B^{-1}$.

Solving Systems of Linear Equations Using Inverse Matrices

For questions 333 and 334, solve using inverse matrices. Refer to the following discussion, as needed.

A solution to the linear system $\begin{array}{l} a_1x_1 + b_1x_2 = c_1 \\ a_2x_1 + b_2x_2 = c_2 \end{array}$ is an ordered pair of numbers (s_1, s_2) such that $\begin{array}{l} a_1s_1 + b_1s_2 = c_1 \\ a_2s_1 + b_2s_2 = c_2 \end{array}$ are both true. When the system has a solution, it is consistent; otherwise, the system is inconsistent. By the definition of matrix multiplication, the linear system $\begin{array}{l} a_1x_1 + b_1x_2 = c_1 \\ a_2x_1 + b_2x_2 = c_2 \end{array}$ is equivalent to the matrix equation $\begin{bmatrix} a_1 & b_1 \\ a_2 & b_2 \end{bmatrix}\begin{bmatrix} x_1 \\ x_2 \end{bmatrix} = \begin{bmatrix} c_1 \\ c_2 \end{bmatrix}$. In matrix notation this equation has the form $AX = C$, where A is the 2×2 coefficient matrix $\begin{bmatrix} a_1 & b_1 \\ a_2 & b_2 \end{bmatrix}$, X is the 2×1 column vector of variables $\begin{bmatrix} x_1 \\ x_2 \end{bmatrix}$, and C is the 2×1 column vector of constants $\begin{bmatrix} c_1 \\ c_2 \end{bmatrix}$.

If A^{-1} exists, then the system has the unique solution $X = A^{-1}C$. *Caution:* If A does not have an inverse, then $X = A^{-1}C$ does not apply.

Note: This method of solving systems of linear equations works for any $n \times n$ system of linear equations, provided the coefficient matrix has an inverse. Still, for systems larger than 2×2, the computations can be long and tedious, so using technological aids to perform the calculations is your best strategy.

---

333. $\begin{array}{l} 3x_1 + 2x_2 = 4 \\ x_1 - 2x_2 = 3 \end{array}$

334. $\begin{array}{l} 5x_1 - x_2 = 3 \\ 2x_1 + 2x_2 = 2 \end{array}$

Solving Systems of Linear Equations Using Cramer's Rule

Overview of Cramer's rule:

The 2×2 system of linear equations $\begin{aligned} a_1x_1 + b_1x_2 &= k_1 \\ a_2x_1 + b_2x_2 &= k_2 \end{aligned}$, where $A = \begin{bmatrix} a_1 & b_1 \\ a_2 & b_2 \end{bmatrix}$ is the

coefficient matrix, $X = \begin{bmatrix} x_1 \\ x_2 \end{bmatrix}$ is the column vector of variables, and $K = \begin{bmatrix} k_1 \\ k_2 \end{bmatrix}$ is the

column vector of constants, has the unique solution: $x_1 = \dfrac{\begin{vmatrix} k_1 & b_1 \\ k_2 & b_2 \end{vmatrix}}{\begin{vmatrix} a_1 & b_1 \\ a_2 & b_2 \end{vmatrix}}$, $x_2 = \dfrac{\begin{vmatrix} a_1 & k_1 \\ a_2 & k_2 \end{vmatrix}}{\begin{vmatrix} a_1 & b_1 \\ a_2 & b_2 \end{vmatrix}}$,

provided that $\begin{vmatrix} a_1 & b_1 \\ a_2 & b_2 \end{vmatrix} \neq 0$. Let X_i be the determinant of the matrix obtained by

replacing the ith column of A with the column vector K and let D be $\begin{vmatrix} a_1 & b_1 \\ a_2 & b_2 \end{vmatrix}$, the

determinant of the coefficient matrix, then the solution can be written as

$$x_1 = \frac{X_1}{D}, \; x_2 = \frac{X_2}{D}.$$

Similarly, Cramer's rule applied to a 3×3 system of linear equations
$$a_1x_1 + b_1x_2 + c_1x_3 = k_1$$
$$a_2x_1 + b_2x_2 + c_2x_3 = k_2 \text{ yields the unique solution:}$$
$$a_3x_1 + b_3x_2 + c_3x_3 = k_3$$

$$x_1 = \frac{\begin{vmatrix} k_1 & b_1 & c_1 \\ k_2 & b_2 & c_2 \\ k_3 & b_3 & c_3 \end{vmatrix}}{\begin{vmatrix} a_1 & b_1 & c_1 \\ a_2 & b_2 & c_2 \\ a_3 & b_3 & c_3 \end{vmatrix}}, \; x_2 = \frac{\begin{vmatrix} a_1 & k_1 & c_1 \\ a_2 & k_2 & c_2 \\ a_3 & k_3 & c_3 \end{vmatrix}}{\begin{vmatrix} a_1 & b_1 & c_1 \\ a_2 & b_2 & c_2 \\ a_3 & b_3 & c_3 \end{vmatrix}}, \; x_3 = \frac{\begin{vmatrix} a_1 & b_1 & k_1 \\ a_2 & b_2 & k_2 \\ a_3 & b_3 & k_3 \end{vmatrix}}{\begin{vmatrix} a_1 & b_1 & c_1 \\ a_2 & b_2 & c_2 \\ a_3 & b_3 & c_3 \end{vmatrix}},$$

provided $D = \begin{vmatrix} a_1 & b_1 & c_1 \\ a_2 & b_2 & c_2 \\ a_3 & b_3 & c_3 \end{vmatrix} \neq 0$. This solution can be written as

$$x_1 = \frac{X_1}{D}, \; x_2 = \frac{X_2}{D}, \; x_3 = \frac{X_3}{D}, \text{ where } X_i \text{ is the determinant of the matrix}$$

obtained by replacing the ith column of the system's coefficient matrix with the

column vector $K = \begin{bmatrix} k_1 \\ k_2 \\ k_3 \end{bmatrix}$. *Note:* It is useful to know that if $D = 0$ and at least one

$X_i \neq 0$, then there is no solution to the system. If $D = 0$ and all the X_i's equal zero, then there are infinitely many solutions to the system.

For questions 335 to 339, solve the system using Cramer's rule.

335. $\begin{aligned} 2x_1 - 3x_2 &= 16 \\ 5x_1 - 2x_2 &= -4 \end{aligned}$

336. $\begin{aligned} x_1 - 2x_2 &= 7 \\ 3x_1 + 2x_2 &= 5 \end{aligned}$

337. $\begin{aligned} x_1 - 2x_2 - 3x_3 &= -20 \\ 2x_1 + 4x_2 - 5x_3 &= 11 \\ 3x_1 + 7x_2 - 4x_3 &= 33 \end{aligned}$

338. $\begin{aligned} x_1 + 2x_2 - x_3 &= 7 \\ 4x_1 + 3x_2 + 2x_3 &= 1 \\ 9x_1 + 8x_2 + 3x_3 &= 4 \end{aligned}$

339. $\begin{aligned} 2x_1 - 4x_2 + 7x_3 &= 5 \\ 3x_1 + 2x_2 - x_3 &= 2 \\ x_1 - 10x_2 + 15x_3 &= 8 \end{aligned}$

Solving Systems of Linear Equations Using Elementary Row Transformations

Overview of elementary row transformations:

1. Two matrices are equivalent if one is obtained from the other by interchanging any two rows.
2. Two matrices are equivalent if one is obtained from the other by multiplying each element of a row by the same nonzero constant.
3. Two matrices are equivalent if one is obtained from the other by adding a nonzero constant multiple of the elements of one row to the corresponding elements of another row.

Caution: Equivalence is not to be confused with equality. The symbol ~ is used to indicate equivalence.

Row transformations enable you to transform a matrix representing a given system of equations to an equivalent matrix from which the solution to the system of equations is obvious or at least simpler to obtain. For the system that has matrix equation $\begin{bmatrix} a_1 & b_1 \\ a_2 & b_2 \end{bmatrix}\begin{bmatrix} x_1 \\ x_2 \end{bmatrix} = \begin{bmatrix} c_1 \\ c_2 \end{bmatrix}$, the matrix representing the system is $\begin{bmatrix} a_1 & b_1 & c_1 \\ a_2 & b_2 & c_2 \end{bmatrix}$ and is called its *augmented matrix*.

Note: The augmented matrices for larger systems are constructed in a similar fashion.

The row transformation procedure that is commonly used is called Gauss-Jordan elimination. In this procedure, the goal is to transform the augmented matrix to reduced row echelon form (also called row canonical form). A matrix is in reduced row echelon form if (a) the first nonzero element in any row is 1 and this element appears to the right of the first nonzero element of the preceding row, and (b) the first nonzero element in a given row is the only nonzero element in its column.

For questions 340 to 342, solve the system of linear equations using elementary row transformations.

340. $\begin{aligned} 3x_1 + 2x_2 &= 4 \\ x_1 - 2x_2 &= 3 \end{aligned}$

341. $\begin{aligned} x_1 - 3x_2 + x_3 &= 2 \\ 2x_1 - x_2 - 2x_3 &= 1 \\ 3x_1 + 2x_2 - x_3 &= 5 \end{aligned}$

342. $\begin{aligned} 2x_1 - x_2 &= 0 \\ 2x_2 - x_3 &= 0 \\ x_1 + 2x_2 - x_3 &= 3 \end{aligned}$

CHAPTER 8

Sequences, Series, and Mathematical Induction

Sequences

For questions 343 to 345, write the closed formula for the arithmetic sequence.

343. 2, 5, 8, 11, 14, …

344. 6, 4, 2, 0, …

345. $\dfrac{7}{3}, \dfrac{16}{3}, \dfrac{25}{3}, \dfrac{34}{3}, \dots$

For questions 346 and 347, write the recursive formula for the arithmetic sequence.

346. 2, 5, 8, 11, 14, …

347. 6, 4, 2, 0, …

For questions 348 to 350, write the closed formula for the geometric sequence.

348. 2, 6, 18, 54, …

349. $\dfrac{1}{3}, \dfrac{2}{15}, \dfrac{4}{75}, \dfrac{8}{375}, \dots$

350. $3, -6, 12, -24, \dots$

For questions 351 and 352, write the recursive formula for the geometric sequence.

351. 2, 6, 18, 54, …

352. $\dfrac{1}{3}, \dfrac{2}{15}, \dfrac{4}{75}, \dfrac{8}{375}, \dots$

353. Given $a_n = 3n + 2$:

    (A) Generate the first four terms of the sequence.

    (B) Find a_{17}.

354. Given $a_n = \dfrac{1}{2}n - 1$:

    (A) Generate the first four terms of the sequence.

    (B) Find a_{19}.

355. Given $a_n = -2n - 1$:

    (A) Generate the first four terms of the sequence.

    (B) Find a_{100}.

356. Given $a_n = 2\left(\dfrac{1}{3}\right)^{n-1}$:

    (A) Generate the first four terms of the sequence.

    (B) Find a_7.

357. Given $a_n = 3\left(-\dfrac{1}{2}\right)^{n-1}$:

    (A) Generate the first four terms of the sequence.

    (B) Find a_7.

For questions 358 to 360, determine whether the given sequence is arithmetic, geometric, both, or neither.

358. 1, 0.5, 0.25, 0.125, …

359. $\dfrac{1}{4}, \dfrac{3}{4}, \dfrac{3}{8}, 1, \ldots$

360. 2, 2, 2, 2, …

361. How many terms are included in the list 2, 7, 12, 17, …, 147?

362. How many terms are included in the list 2, 4, 8, 16, …, 4096?

Series

In questions 363 to 367, use the sequence whose formula is $a_n = 5n - 2$. Refer to the following properties of the summation (sigma) notation, as needed.

(a)	$\displaystyle\sum_{k=1}^{n} c = nc$	A constant, c, added n times is nc.
(b)	$\displaystyle\sum_{k=1}^{n} (a_k \pm b_k) = \sum_{k=1}^{n} a_k \pm \sum_{k=1}^{n} b_k$	The terms can be summed or subtracted individually.
(c)	$\displaystyle\sum_{k=1}^{n} c a_k = c \sum_{k=1}^{n} a_k$	A constant can be factored out of a summation.
(d)	$\displaystyle\sum_{k=1}^{n} a_k = \sum_{k=1}^{m} a_k + \sum_{k=m+1}^{n} a_k$	Parts can be summed individually.
(e)	$\displaystyle\sum_{k=1}^{n} a_k = \sum_{j=1}^{n} a_j = \sum_{m=0}^{n-1} a_{m+1}$	The subscripting can be adjusted to fit special purposes.
(f)	$\displaystyle\sum_{k=1}^{n} a_k = a_1 + a_2 + \sum_{k=3}^{n-1} a_k + a_n$	Terms can be "taken out" of the sum if needed.

363. Write the sum of the first 15 terms, using the sigma notation where the first two terms have been taken out of the sum.

364. Write the sum of the first n terms, using the sigma notation such that the summing index begins with 2.

365. Write the sum of the first 15 terms and apply summation notation properties (a), (b), and (c) to rewrite the sum.

366. Write the extended form of the sum $\displaystyle\sum_{k=1}^{n} (5k - 2)$.

367. Compute the sum $\displaystyle\sum_{k=1}^{6} (5k - 2)$.

For questions 368 to 372, compute the indicated sum. Refer to the following special closed formulas, as needed.

$$S_n = \sum_{k=1}^{n} k = \frac{n(n+1)}{2}$$

$$S_n = \sum_{k=1}^{n} [a_1 + (k-1)d)] = \frac{n(a_1 + a_n)}{2}$$ where a_1 is the first term and a_n is the last term of the arithmetic sequence

$$S_n = \sum_{k=1}^{n} a_1 r^{k-1} = \frac{a_1(1-r^n)}{1-r},$$ provided $r \neq 1$, where a_1 is the first term and r is the common ratio of the geometric sequence

368. $S_7 = \displaystyle\sum_{k=1}^{7} \left(\frac{1}{2}\right)^{k-1}$

369. $\displaystyle\sum_{k=1}^{12} [3 + (k-1)2]$

370. $4 + 7 + 10 + 13 + 16 + 19 + 22 + 25$

371. $1 + 2 + 4 + 8 + 16 + 32 + 64 + 128$

372. $3 + 5 + 7 + 9 + \ldots + 477$

Mathematical Induction

For questions 373 to 382, use mathematical induction to prove the given statement is true for all natural numbers n. Refer to the following statement of the principle of mathematical induction, as needed.

The Principle of Mathematical Induction

Let S_n be a statement involving natural numbers, where $n = 1, 2, 3, \ldots,$ $k, k+1, \ldots$. If

1. S_1 is true, and
2. If S_k is true implies S_{k+1} is true

then S_n is true for all natural numbers n.

Note: A proof by mathematical induction is a two-part proof. The statement "If S_k is true" is termed the *induction hypothesis* (IH). Mathematical induction is a technique for proving that a statement involving natural numbers is true. In many instances, you observe a pattern involving natural numbers, and then you use induction to verify that the pattern you observed is indeed true for all natural numbers.

373. $S_n : 2 + 4 + 6 + \ldots + 2n = n(n+1)$

374. $5 + 9 + 13 + \ldots + (4n+1) = n(2n+3)$

375. $2 + 4 + 8 + \ldots + 2^n = 2^{n+1} - 2$

376. $\dfrac{1}{3} + \dfrac{1}{15} + \dfrac{1}{35} + \ldots + \dfrac{1}{(2n-1)(2n+1)} = \dfrac{n}{2n+1}$

377. $3 + 9 + 27 + \ldots + 3^n = \dfrac{3(3^n - 1)}{2}$

378. $2^{n+1} \geq 2n + 1$

379. $\dfrac{1}{2} + \dfrac{1}{6} + \dfrac{1}{12} + \ldots + \dfrac{1}{n(n+1)} = \dfrac{n}{n+1}$

380. $a + ar + ar^2 + ar^3 + \ldots + ar^{n-1} = \dfrac{a(1-r^n)}{1-r}$ provided $r \neq 1$

381. $1 + 8 + 27 + 64 + \ldots + n^3 = \dfrac{n^2(n+1)^2}{4}$

382. $4^n - 1 \geq 3(4^{n-1})$

CHAPTER **9**

Trigonometric Functions

Angle Measurement

For questions 383 to 392, convert the angle measurements as indicated. Recall that because $2\pi = 360°$, $1° = \dfrac{\pi}{180}$ radians and 1 radian $= \dfrac{180°}{\pi}$.

383. Convert $45°$ to radians.

384. Convert 1 radian to degrees.

385. Convert $60°$ to radians.

386. Convert $\dfrac{\pi}{6}$ radians to degrees.

387. Convert $127°$ to radians.

388. Convert 25 radians to degrees.

389. Convert $135°$ to radians.

390. Convert $0°$ to radians.

391. Convert $28°\,43'\,25''$ to decimal form. Round to two decimal places.

392. Convert 57.5692 to degree, minute, second form. Round seconds to the nearest whole number.

Trigonometric Functions

Important concepts to recall:

1. Definitions of trigonometric (trig) functions: If an angle θ in the x-y plane is in standard position (that is, its vertex is at the origin and its initial side coincides with the positive x-axis) and if (a,b) is on the terminal side of θ and r is the distance from $(0,0)$ to (a,b) as shown in the figure below, then the trig functions have the following definitions:

$$\sin\theta = \frac{b}{r}, \cos\theta = \frac{a}{r}, \tan\theta = \frac{b}{a}, \csc\theta = \frac{r}{b}, \sec\theta = \frac{r}{a}, \text{ and } \cot\theta = \frac{a}{b}.$$

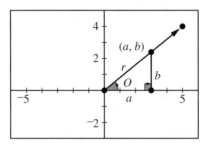

Note: From the preceding definitions, sine and cosecant are reciprocals of each other, cosine and secant are reciprocals of each other, and tangent and cotangent are reciprocals of each other. Therefore, it is necessary to remember the definitions for the sine, cosine, and tangent functions only because you can determine the other functions by using the reciprocal relationships.

2. Unit circle: As shown in the figure below, the unit circle is centered at the origin with radius $r = 1$. If (x, y) is a point on the unit circle, lying on the terminal side of an angle θ in standard position, then the y and x coordinates are the sine and cosine of angle θ, respectively. As the point (x, y) rotates counterclockwise around the circle, values of the trig functions can be determined. For instance, at $\theta = 0$, $\sin\theta = \sin 0 = \dfrac{y}{r} = \dfrac{0}{1} = 0$ and $\cos\theta = \cos 0 = \dfrac{x}{r} = \dfrac{1}{1} = 1$; and at $\theta = \dfrac{\pi}{2}$, $\sin\theta = \sin\dfrac{\pi}{2} = \dfrac{y}{r} = \dfrac{1}{1} = 1$ and $\cos\theta = \cos\dfrac{\pi}{2} = \dfrac{x}{r} = \dfrac{0}{1} = 0$. Furthermore, if the terminal side of θ does *not* lie on an axis, a trig function of θ is positive or negative depending on the quadrant in which the terminal side of θ is located. You need to remember only in which quadrants the sine, cosine, and tangent are positive because they are negative in the other quadrants, and their reciprocals have the same signs as the functions themselves do. The sine function is positive in

quadrants I and II; the cosine function is positive in quadrants I and IV; and the tangent function is positive in quadrants I and III.

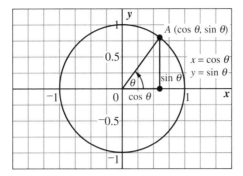

3. Reference angles: A reference angle for an angle A in standard position (whose terminal side does *not* lie on an axis) is the positive angle between the x-axis and the terminal side of the angle. The following figure depicts $\angle A$ and its associated reference angle, $\angle A_r$, as the terminal side of $\angle A$ rotates counterclockwise around the origin from quadrant I (a) to quadrant II (b) to quadrant III (c) to quadrant IV (d).

   Angles whose terminal sides coincide with a coordinate axis are their own reference angles.

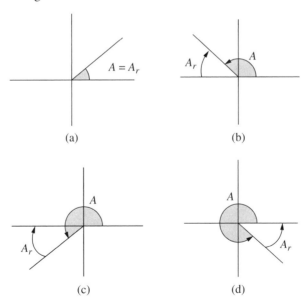

4. Right triangle trig ratios: The ratios relative to angle A in right triangle ABC are as follows:

$$\text{sine of } \angle A = \sin A = \frac{\text{side opposite}}{\text{hypotenuse}} = \frac{a}{c}$$

$$\text{cosine of } \angle A = \cos A = \frac{\text{side adjacent}}{\text{hypotenuse}} = \frac{b}{c}$$

$$\text{tangent of } \angle A = \tan A = \frac{\text{side opposite}}{\text{side adjacent}} = \frac{a}{b}$$

$$\text{cosecant of } \angle A = \csc A = \frac{\text{hypotenuse}}{\text{side opposite}} = \frac{c}{a}$$

$$\text{secant of } \angle A = \sec A = \frac{\text{hypotenuse}}{\text{side adjacent}} = \frac{c}{b}$$

$$\text{cotangent of } \angle A = \cot A = \frac{\text{side adjacent}}{\text{side opposite}} = \frac{b}{a}$$

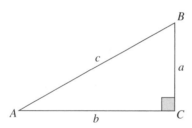

5. Special right triangles: Two special right triangles from which you can calculate trigonometric ratios are shown below.

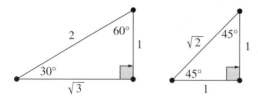

Note: Any triangles similar to these two right triangles have the same trig ratio values.

For questions 393 to 407, find the exact value of the trigonometric function. Do not use a calculator.

393. $\cos 60°$

394. $\sin 30°$

395. $\tan 45°$

396. $\cos \dfrac{3\pi}{2}$

397. $\tan 60°$

398. $\sin \dfrac{3\pi}{2}$

399. $\cos(\pi)$

400. $\sin 45°$

401. $\cos 45°$

402. $\tan \dfrac{\pi}{2}$

403. $\cos 135°$

404. $\tan 210°$

405. $\cos 2\pi$

406. $\tan \pi$

407. $\sin \pi$

Trigonometric Graphs and Transformations

Important concepts to recall (continued):

6. Graphs of the three main trig functions: As shown in the following graphs, the sine, cosine, and tangent functions are periodic functions. Recall that $y = \sin x$ and $y = \cos x$ have period 2π, amplitude 1, maximum value 1, and minimum value -1; and $y = \tan x$ has period π, no amplitude, and no maximum or minimum.

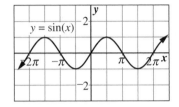 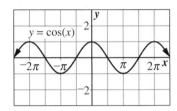

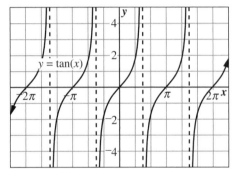

7. The functions $y = A\sin(Bx + C) + K$ and $y = A\cos(Bx + C) + K$ are transformations of the basic sine and cosine functions. The constant A induces a vertical compression or stretch; and if $A < 0$, a reflection about the x-axis. The constant B induces a horizontal compression or stretch, the constant C induces a horizontal shift (usually called a *phase* shift), and the constant K induces a vertical shift. Therefore, the functions $y = A\sin(Bx + C) + K$ and $y = A\cos(Bx + C) + K$ have amplitude $|A|$, period $\dfrac{2\pi}{|B|}$, horizontal shift $\left|\dfrac{C}{B}\right|$ — to the right if $\dfrac{C}{B} < 0$ and to the left if $\dfrac{C}{B} > 0$, and vertical shift $|K|$ — up if $K > 0$ and down if $K < 0$. The maximum height of the graph is $|A| + K$ and the minimum height of the graph is $-|A| + K$. *Note:* To more easily determine a horizontal shift, rewrite $(Bx + C)$ as $B\left(x + \dfrac{C}{B}\right)$ because the horizontal shift is actually $\left|\dfrac{C}{B}\right|$.

8. The function $y = A\tan(Bx + C) + K$ is a transformation of the basic tangent function. The constant A induces a vertical compression or stretch; and if $A < 0$, a reflection about the x-axis. The constant B induces a horizontal compression or stretch, the constant C induces a phase shift, and the constant K induces a vertical shift. Therefore, the function

$y = A\tan(Bx + C) + K$ has period $\dfrac{\pi}{|B|}$, horizontal shift $\left|\dfrac{C}{B}\right|$—to the right

if $\dfrac{C}{B} < 0$ and to the left if $\dfrac{C}{B} > 0$, and vertical shift $|K|$—up if $K > 0$ and

down if $K < 0$. However, unlike the sine and cosine functions, the tangent function has neither a maximum nor a minimum value.

For questions 408 to 412, find the period of the function.

408. $y = 3\sin x$

409. $y = 2\cos 3x + 5$

410. $y = \sin(-5x + 2)$

411. $y = 4\cos\left(\dfrac{1}{2}x - \pi\right)$

412. $y = -3\sin(x + 2\pi) - 1$

For questions 413 to 417, determine the (a) phase shift and (b) vertical shift.

413. $y = \sin(x + 2\pi) + 4$

414. $y = 3\cos\left(2x - \dfrac{\pi}{3}\right)$

415. $y = 2\sin\left(3x + \dfrac{\pi}{5}\right) - 1$

416. $y = -2\cos\left(\dfrac{1}{3}x - \pi\right)$

417. $y = 4\sin(2x - \sqrt{2}) + \dfrac{3}{5}$

For questions 418 to 422, find the amplitude A of the function.

418. $y = 3\sin x$

419. $y = -4\cos x - 3$

420. $y = \dfrac{1}{2}\sin(5x - 2\pi)$

421. $y = \pi\cos(\pi x + 3\pi) + 3$

422. $y = -12\cos(22x - 17) + 25$

For questions 423 to 427, determine the (a) period, (b) phase shift, and (c) vertical shift.

423. $y = 3\tan(2x)$

424. $y = \tan(-3\theta + 2\pi) - 5$

425. $y = 3\tan(\pi x - 2) + 10$

426. $y = 7\tan\left(\dfrac{3}{\pi}x + 5\right)$

427. $y + 2 = \cos(3x - 6)$

428. Using the following figure, solve $\sin x = 0$ when $-2\pi \le x \le 2\pi$.

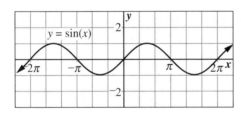

429. Using the following figure, solve $\cos x = 0$ when $-2\pi \le x \le 2\pi$.

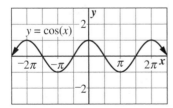

For questions 430 to 432, sketch the graph of the function.

430. $y = 3\sin(2x - \pi)$

431. $y = -2\cos(-3x - 6\pi)$

432. $y = \dfrac{1}{2}\tan(3x - 6)$

CHAPTER 10

Analytic Trigonometry

Fundamental Trigonometric Identities

For questions 433 to 442, prove that the given equation is an identity. Refer to the following table of fundamental identities, as needed.

Fundamental Trigonometric Identities

Reciprocal Identities	Ratio Identities	Pythagorean Identities	Even-Odd Function Identities
$\sin\theta = \dfrac{1}{\csc\theta}$	$\tan\theta = \dfrac{\sin\theta}{\cos\theta}$	$\sin^2\theta + \cos^2\theta = 1$	$\sin(-\theta) = -\sin\theta$
$\csc\theta = \dfrac{1}{\sin\theta}$		$1 - \sin^2\theta = \cos^2\theta$	
		$1 - \cos^2\theta = \sin^2\theta$	
$\cos\theta = \dfrac{1}{\sec\theta}$	$\cot\theta = \dfrac{\cos\theta}{\sin\theta}$	$1 + \tan^2\theta = \sec^2\theta$	$\cos(-\theta) = \cos\theta$
		$\tan^2\theta = \sec^2\theta - 1$	
$\sec\theta = \dfrac{1}{\cos\theta}$		$\sec^2\theta - \tan^2\theta = 1$	
$\tan\theta = \dfrac{1}{\cot\theta}$	$\tan\theta = \dfrac{\sec\theta}{\csc\theta}$	$1 + \cot^2\theta = \csc^2\theta$	$\tan(-\theta) = -\tan\theta$
		$\cot^2\theta = \csc^2\theta - 1$	
$\cot\theta = \dfrac{1}{\tan\theta}$	$\cot\theta = \dfrac{\csc\theta}{\sec\theta}$	$\csc^2\theta - \cot^2\theta = 1$	

Note: The convenient notation for exponents on trig functions, $(\sin\theta)^2 = \sin^2\theta$, will be used throughout this book.

A word of advice: Not only should you memorize the identities in the preceding table but also *own* them. That is, they should be an instinctive part of your mathematical knowledge.

433. $\sin x \cot x = \cos x$

434. $\sin x (\csc x - \sin x) = \cos^2 x$

435. $(1 + \sin x)(1 + \sin(-x)) = \cos^2 x$

436. $\tan^2 x \csc^2 x - \tan^2 x = 1$

437. $\tan x (\csc x + \cot x) = \sec x + 1$

438. $\dfrac{\sin x \cos x + \cos x}{\sin x + \sin^2 x} = \cot x$

439. $\dfrac{\cos^2 x}{\sin x} + \dfrac{\sin x}{1} = \csc x$

440. $\dfrac{\sin x + \cos x}{\tan x} = \cos x + \dfrac{\cos^2 x}{\sin x}$

441. $(\sin x + \cos x)^2 = 1 + 2\sin x \cos x$

442. $\sec^4 x - \tan^4 x = \sec^2 x + \tan^2 x$

Sum and Difference Identities

For questions 443 to 455, prove that the given equation is an identity. Refer to the following sum and difference identities, as needed.

$$\sin(\theta \pm \beta) = \sin\theta \cos\beta \pm \cos\theta \sin\beta$$

$$\cos(\theta \pm \beta) = \cos\theta \cos\beta \mp \sin\theta \sin\beta$$

$$\tan(\theta \pm \beta) = \frac{\tan\theta \pm \tan\beta}{1 \mp \tan\theta \tan\beta}$$

443. $\sin(x + 2\pi) = \sin x$

444. $\cos(x + 2\pi) = \cos x$

445. $\tan(x + \pi) = \tan x$

446. $\cos\left(x + \dfrac{\pi}{4}\right) = \dfrac{\sqrt{2}}{2}(\cos x - \sin x)$

447. $\sin(\alpha + \beta)\sin(\alpha - \beta) = \sin^2 \alpha - \sin^2 \beta$

448. $\tan\left(x + \dfrac{\pi}{4}\right) = \dfrac{1 + \tan x}{1 - \tan x}$

449. $\sin(2\theta) = 2 \sin \theta \cos \theta$

450. $\cos(2\theta) = 2 \cos^2 \theta - 1$

451. $\sin(4x) = 4 \sin x \cos x(1 - 2 \sin^2 x)$

452. $\dfrac{\cos(2x)}{\sin^2 x} = \cot^2 x - 1$

453. $\dfrac{\cos(2x)}{\sin(2x)} = \dfrac{\cot x - \tan x}{2}$

454. $\tan(2\theta) = \dfrac{2 \tan \theta}{1 - \tan^2 \theta}$

455. $\tan(2x) = \dfrac{2 \sin x \cos x}{\cos^2 x - \sin^2 x}$

Double- and Half-Angle Identities

For questions 456 to 458, prove the given equation is an identity. Refer to the following double- and half-angle identities, as needed.

$$\sin(2\theta) = 2 \sin \theta \cos \theta$$

$$\cos(2\theta) = \cos^2 \theta - \sin^2 \theta = 2 \cos^2 \theta - 1 = 1 - 2 \sin^2 \theta$$

$$\tan(2\theta) = \dfrac{2 \tan \theta}{1 - \tan^2 \theta}$$

$$\sin\left(\dfrac{\theta}{2}\right) = \pm\sqrt{\dfrac{1 - \cos \theta}{2}}$$

$$\cos\left(\dfrac{\theta}{2}\right) = \pm\sqrt{\dfrac{1 + \cos \theta}{2}}$$

$$\tan\left(\dfrac{\theta}{2}\right) = \dfrac{\sin \theta}{1 + \cos \theta} = \dfrac{1 - \cos \theta}{\sin \theta}$$

456. $\cot\left(\dfrac{x}{2}\right) = \dfrac{1 + \cos x}{\sin x}$

457. $\cot(4x) = \dfrac{1 - \tan^2 2x}{2 \tan 2x}$

458. $\sin^2\left(\dfrac{x}{2}\right) = \dfrac{\sec x - 1}{2 \sec x}$

459. Verify that $\cos(2\theta) = 2\cos\theta$ is not an identity.

460. Verify that $\sin(\theta + \beta) = \sin\theta + \sin\beta$ is not an identity.

Trigonometric Equations

For questions 461 and 462, determine whether the given number is a solution to the given equation.

461. $\dfrac{\pi}{3}$, $2\sin x = \sqrt{3}$

462. $\dfrac{\pi}{3}$, $2\sec x = \tan x + \cot x$

For questions 463 to 467, solve the equation for x in the given interval. Refer to the following table of special reference angles, as needed. (See Chapter 9 for a discussion of reference angles.)

Trigonometric Function Values of Special Angles

Angle (°)	Angle (Radians)	Sine	Cosine	Tangent	Cotangent	Secant	Cosecant
0	0	0	1	0	undefined	1	undefined
30°	$\dfrac{\pi}{6}$	$\dfrac{1}{2}$	$\dfrac{\sqrt{3}}{2}$	$\dfrac{1}{\sqrt{3}}$	$\sqrt{3}$	$\dfrac{2}{\sqrt{3}}$	2
45°	$\dfrac{\pi}{4}$	$\dfrac{1}{\sqrt{2}}$	$\dfrac{1}{\sqrt{2}}$	1	1	$\sqrt{2}$	$\sqrt{2}$
60°	$\dfrac{\pi}{3}$	$\dfrac{\sqrt{3}}{2}$	$\dfrac{1}{2}$	$\sqrt{3}$	$\dfrac{1}{\sqrt{3}}$	2	$\dfrac{2}{\sqrt{3}}$
90°	$\dfrac{\pi}{2}$	1	0	undefined	0	undefined	1

463. $2 \sin x - 1 = 0, [0, 2\pi)$

464. $\sin x + \cos x = 0, (-\infty, \infty)$

465. $\sin x \tan x = \sin x, [0, 2\pi)$

466. $4 \cos^2 x - 3 = 0, [0, 2\pi)$

467. $\cos^3 x + \cos x = 0, [0, 2\pi)$

For questions 468 to 472, solve the equation for x in the given interval. Use inverse functions and refer to the following table, as needed. Use a calculator and round answers to two decimal places when necessary.

Inverse Trigonometric Functions and Their Restricted Domains

Inverse Trigonometic Functions	Restricted Domains
$y = \sin \theta$ if and only if $\theta = \sin^{-1} y$	$-\dfrac{\pi}{2} \leq \theta \leq \dfrac{\pi}{2}$ or $-90° \leq \theta \leq 90°$
$y = \cos \theta$ if and only if $\theta = \cos^{-1} y$	$0 \leq \theta \leq \pi$ or $0° \leq \theta \leq 180°$
$y = \tan \theta$ if and only if $\theta = \tan^{-1} y$	$-\dfrac{\pi}{2} < \theta < \dfrac{\pi}{2}$ or $-90° < \theta < 90°$

Note: It is convenient to read $\theta = \sin^{-1} y$ as "θ is the angle whose sine is y" and similarly, for the other inverse functions.

468. $\tan x + 3 \cot x = 4, [0°, 360°)$

469. $\cos x = 1 + \sqrt{3} \sin x, [0, 2\pi)$

470. $\cos\left(\dfrac{x}{2}\right) = \dfrac{1}{2}, [0, 2\pi)$

471. $\sin(2x) = \cos(2x), [0, 2\pi)$

472. $2 \cos x + \sin x = 1, [0°, 360°)$

Solving Triangles

For questions 473 to 477, solve right triangle *ABC*, where $\angle C = 90°$. Round all final measures to the nearest tenth. *Note:* Avoid rounding intermediate answers. Instead use the "recall answer" feature of your calculator to use a value in a subsequent calculation.

473.

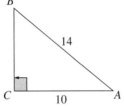

474.

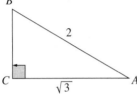

475.

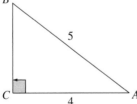

476.

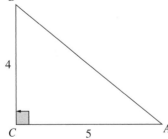

477.

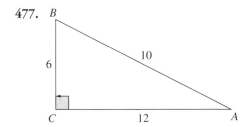

For questions 478 to 482, solve oblique triangle *ABC*. Round all final measures to the nearest tenth. Refer to the following law of sines and law of cosines, as needed.

Law of Sines and Law of Cosines

Law of Sines	Law of Cosines	Oblique Triangle
$\dfrac{\sin A}{a} = \dfrac{\sin B}{b} = \dfrac{\sin C}{c}$	$a^2 = b^2 + c^2 - 2bc \cos A$ $b^2 = a^2 + c^2 - 2ac \cos B$ $c^2 = a^2 + b^2 - 2ab \cos C$	

478. Given $a = 75$, $\angle A = 38°$, and $\angle B = 64°$

479. $c = 126$, $\angle A = 13°$, and $\angle B = 22°$

480. $a = 42.7$, $\angle B = 103.4°$, and $\angle C = 19.6°$

481. $b = 3.2$, $c = 1.5$, and $\angle A = 95.7°$

482. $a = 75$, $b = 32$, and $\angle C = 38°$

CHAPTER 11

Conic Sections

The Circle

For questions 483 to 487, find the center and radius of the given circle. Refer to the following guidelines, as needed.

> A circle with center (h,k) and radius r has standard equation
> $(x-h)^2 + (y-k)^2 = r^2$.

483. $x^2 + y^2 = 1$

484. $x^2 - 4x + 2y + y^2 = 4$

485. $x^2 + y^2 - 4x + 10y = -4$

486. $3x^2 - 12y + 6x + 3y^2 = 49$

487. Write an equation for the circle centered at $(-4,-7)$ with radius $= \sqrt{11}$.

488. Write an equation for the circle that passes through the points $(0,0)$, $(0,2)$, and $(2,0)$.

The Ellipse

For questions 489 to 493, find the center, vertices, and lengths of the major and minor axes for each given ellipse. Refer to the following guidelines, as needed.

An ellipse with horizontal major axis and center (h,k) has standard equation

$$\frac{(x-h)^2}{a^2} + \frac{(y-k)^2}{b^2} = 1 \qquad (a > b > 0)$$

with vertices $(h \pm a, k)$ and y-intercepts $(h, k \pm b)$. The line segment joining the vertices $(h \pm a, k)$ is the major axis and has length $2a$, and the line segment joining the y-intercepts $(h, k \pm b)$ is the minor axis and has length $2b$.

An ellipse with vertical major axis and center (h,k) has standard equation

$$\frac{(x-h)^2}{b^2} + \frac{(y-k)^2}{a^2} = 1 \qquad (a > b > 0)$$

with vertices $(h, k \pm a)$ and x-intercepts $(h \pm b, k)$. The line segment joining the vertices $(h, k \pm a)$ is the major axis and has length $2a$, and the line segment joining the x-intercepts $(h \pm b, k)$ is the minor axis and has length $2b$.

---

489. $4x^2 + 25y^2 = 100$

490. $9x^2 + 4y^2 = 36$

491. $x^2 + 4y^2 - 8y + 4x - 8 = 0$

492. $5x^2 + 2y^2 + 20y - 30x + 75 = 0$

493. $2x^2 + 5y^2 - 12x + 20y - 12 = 0$

The Hyperbola

For questions 494 to 497, find the center, vertices, equations of the asymptotes, and lengths of the tranverse and conjugate axes for each given hyperbola. Refer to the following guidelines, as needed.

A hyperbola with horizontal transverse axis and center (h,k) has standard equation

$$\frac{(x-h)^2}{a^2} - \frac{(y-k)^2}{b^2} = 1 \qquad (a,b>0)$$

with vertices $(h \pm a, k)$ and asymptotes $y = \pm \frac{b}{a}(x-h)+k$. The transverse axis has length $2a$, and the conjugate axis has length $2b$.

A hyperbola with vertical transverse axis and center (h,k) has standard equation

$$\frac{(y-k)^2}{a^2} - \frac{(x-h)^2}{b^2} = 1 \qquad (a,b>0)$$

with vertices $(h, k \pm a)$ and asymptotes $y = \pm \frac{a}{b}(x-h)+k$. The transverse axis has length $2a$, and the conjugate axis has length $2b$.

494. $36x^2 - 16y^2 = 144$

495. $16y^2 - 4x^2 + 24x = 100$

496. $9y^2 - x^2 + 54y + 4x + 68 = 0$

497. $4x^2 - 9y^2 - 24x + 72y - 144 = 0$

The Parabola

For questions 498 to 501, find the vertex, focus, and directrix for each given parabola. Tell whether the parabola opens upward or downward or to the right or left. Refer to the following guidelines, as needed.

A parabola with vertical axis has standard equation

$$(x - h)^2 = 4p(y - k)$$

with vertex (h, k), focus $(h, k + p)$, and directrix $y = k - p$. If $p > 0$, the parabola opens upward, and if $p < 0$, the parabola opens downward.
  A parabola with horizontal axis has standard equation

$$(y - k)^2 = 4p(x - h)$$

with vertex (h, k), focus $(h + p, k)$, and directrix $x = h - p$. If $p > 0$, the parabola opens to the right, and if $p < 0$, the parabola opens to the left.

498. $x^2 = -12y$

499. $y^2 = 4x$

500. $y^2 + 2y + 16x = -49$

501. $x^2 - 8y = 6x - 25$

ANSWERS

Chapter 1: Basic Algebraic Skills

1. Recall that the proper subsets of the real numbers are the natural numbers, the whole numbers, the integers, the rational numbers, and the irrational numbers.

- (A) 15 is a natural number, a whole number, an integer, a rational number, and a real number.
- (B) $\sqrt[3]{\dfrac{64}{125}} = \dfrac{4}{5}$ is a rational, real number.
- (C) $-\pi$ is an irrational, real number.
- (D) -100 is an integer, a rational number, and a real number.
- (E) $\sqrt{3}$ is an irrational, real number.

2. In interval notation, a square bracket means the endpoint is included, and a parenthesis means the endpoint is not included. Open intervals do not include the endpoints. Closed intervals include both endpoints. Half-open (or half-closed) intervals include only one endpoint. Finite intervals are bounded intervals. Intervals that extend indefinitely to the right and left are unbounded.

- (A) $(-\infty, -4.5)$, unbounded below, bounded above, open
- (B) $[-12, 28]$, bounded, closed
- (C) $[-3, \infty)$, bounded below, unbounded above, half-open
- (D) $(-\infty, 0)$, unbounded below, bounded above, open
- (E) $(-\infty, \infty)$, unbounded, open

3.
- (A) closure property of multiplication
- (B) commutative property of addition
- (C) multiplicative inverse property
- (D) closure property of addition
- (E) associative property of addition
- (F) distributive property
- (G) additive inverse property
- (H) zero factor property
- (I) distributive property
- (J) associative property of multiplication

4. The absolute value of a real number x is given by $|x| = \begin{cases} x \text{ if } x \geq 0 \\ -x \text{ if } x < 0 \end{cases}$.

(A) 20.9

(B) $7\dfrac{2}{3}$

(C) –80

(D) $\dfrac{\pi}{3}$

(E) $-a$

5. (A) <

(B) >

(C) =

(D) <

(E) <

6. To add two numbers that have the same sign, add their absolute values and give the sum their common sign. Thus, $-100 + -50 = -150$.

7. To add two numbers that have opposite signs, subtract the lesser absolute value from the greater absolute value and give the sum the sign of the number with the greater absolute value. Thus, $0.8 + -1.9 = -1.1$.

8. To subtract two signed numbers, add the opposite of the number that follows the minus symbol. Thus, $-34 - (-22) = -34 + 22 = -12$.

9. To multiply two numbers that have opposite signs, multiply their absolute values and make the product negative. Thus, $\left(-\dfrac{35}{41}\right)\left(\dfrac{2}{5}\right) = -\dfrac{14}{41}$.

10. To divide two numbers, divide their absolute values (being careful to make sure you don't divide by 0) and then follow the rules for multiplication of signed numbers. Thus, $\dfrac{24}{-3} = -8$.

11. To multiply two numbers that have the same sign, multiply their absolute values and keep the product positive. Thus, $(-200)(-6) = 1200$.

12. $-175.54 + 3.48 + 1.23 = -170.83$.

13. $125 - (-437) - 80 + 359 = 841$

14. $(-0.25)(-600)(4.5)(-50) = -33{,}750$

15. $\dfrac{-800}{-0.04} = 20{,}000$

16. If x is any real number and n is a positive integer, then $x^n = \underbrace{x \cdot x \cdot x \cdot \ldots \cdot x}_{n \text{ factors of } x}$. Thus, $(-3)^4 = (-3)(-3)(-3)(-3) = 81$.

17. $-3^4 = -(3 \cdot 3 \cdot 3 \cdot 3) = -81$. Notice that the exponent applies only to the number 3 and not to the negation symbol.

18. For a number x and a natural number n, $x^{\frac{1}{n}} = \sqrt[n]{x}$, provided that when n is even, $x \geq 0$. Thus, $49^{\frac{1}{2}} = \sqrt{49} = 7$.

19. A nonzero number raised to a negative exponent equals the reciprocal of the number raised to the corresponding positive exponent. Thus, $5^{-3} = \dfrac{1}{5^3} = \dfrac{1}{125}$.

20. $x^{\frac{p}{q}} = \left(x^{\frac{1}{q}}\right)^p = (\sqrt[q]{x})^p$ or $x^{\frac{p}{q}} = (x^p)^{\frac{1}{q}} = \sqrt[q]{x^p}$ provided, in all cases, that $x \geq 0$ when q is even and division by zero or 0^0 does not occur. Thus, $32^{\frac{3}{5}} = (\sqrt[5]{32})^3 = 2^3 = 8$.

21. Any nonzero number raised to the zero power is 1. Thus, $(\sqrt{2})^0 = 1$.

22. $x^m x^n = x^{m+n}$. Thus, $x^5 x^3 = x^{5+3} = x^8$.

23. $\dfrac{x^m}{x^n} = x^{m-x}$. Thus, $\dfrac{y^7}{y^2} = y^{7-2} = y^5$.

24. $(x^n)^p = x^{np}$. Thus, $(z^3)^4 = z^{3 \cdot 4} = z^{12}$.

25. $\left(\dfrac{x}{y}\right)^p = \dfrac{x^p}{y^p}$. Thus, $\left(\dfrac{a}{b}\right)^4 = \dfrac{a^4}{b^4}$.

26. $\left(\dfrac{x}{y}\right)^{-n} = \left(\dfrac{y}{x}\right)^n$. Thus, $\left(\dfrac{x}{y}\right)^{-2} = \left(\dfrac{y}{x}\right)^2 = \dfrac{y^2}{x^2}$.

27. $(a^3 b^6)^{\frac{1}{3}} = a^{3 \cdot \frac{1}{3}} b^{6 \cdot \frac{1}{3}} = ab^2$

28. $(u^{-2})^4 v^7 v^{-9} = u^{-8}v^{-2} = \dfrac{1}{u^8 v^2}$

29. $\dfrac{(x^2 y^{-5})^{-4}}{(x^5 y^{-2})^{-3}} = \dfrac{(x^5 y^{-2})^3}{(x^2 y^{-5})^4} = \dfrac{x^{15} y^{-6}}{x^8 y^{-20}} = x^7 y^{14}$

30. $\dfrac{1}{2^{-1} + 3^{-1}} = \dfrac{1}{\dfrac{1}{2} + \dfrac{1}{3}} = \dfrac{1}{\dfrac{3}{6} + \dfrac{2}{6}} = \dfrac{1}{\dfrac{5}{6}} = \dfrac{6}{5}$

31. $(7+8)20 - 10 = (15)20 - 10 = 290$

32. $(-9^2)(5-8) = (-81)(-3) = 243$

33. $(20-(-30))\left(-400^{\frac{1}{2}}\right) = (50)(-20) = -1000$

34. $8(-2) - \dfrac{15}{-5} = -16 - (-3) = -16 + 3 = -13$

35. $109 - \dfrac{20+22}{-6} - 4^3 = 109 - \dfrac{42}{-6} - 64 = 109 - (-7) - 64 = 109 + 7 - 64 = 52$

36. $(-2)^4 \cdot -3 - (15-4)^2 = 16 \cdot -3 - (11)^2 = -48 - 121 = -169$

37. $3(-11 + 3 \cdot 8 - 6 \cdot 3)^2 = 3(-11 + 24 - 18)^2 = 3(-5)^2 = 3(25) = 75$

38. $-15 - \dfrac{-10 - (3 \cdot -3 + 17)}{2} = -15 - \dfrac{-10 - (-9 + 17)}{2} = -15 - \dfrac{-10 - (8)}{2} = -15 - \dfrac{-18}{2}$

$$= -15 - (-9) = -15 + 9 = -6$$

39. $\dfrac{6^2 - 8 \cdot 10 + 3^4 + 2}{3 \cdot 2 - 36 \div 12} = \dfrac{36 - 8 \cdot 10 + 81 + 2}{3 \cdot 2 - 36 \div 12} = \dfrac{36 - 80 + 83}{6 - 3} = \dfrac{39}{3} = 13$

40. $\dfrac{188 - 2(3^2 \cdot 2 - 3 \cdot 2^3)}{10^2} = \dfrac{188 - 2(9 \cdot 2 - 3 \cdot 8)}{100} = \dfrac{188 - 2(18 - 24)}{100} = \dfrac{188 - 2(-6)}{100}$

$$= \dfrac{188 + 12}{100} = \dfrac{200}{100} = 2$$

41. $(x + yi) + (u + vi) = (x + u) + (y + v)i.$ Thus, $(-8 - 5i) + (6 - 7i) = -2 - 12i.$

42. $(9 + 4i) + (5 - 4i) = 14 + 0i = 14$

43. $(1 - i\sqrt{5}) + (-7 + 2i\sqrt{5}) = -6 + i\sqrt{5}$

44. $(x + yi) - (u + vi) = (x - u) + (y - v)i.$ Thus, $(16 - 5i) - (-3 + 7i) = 19 - 12i.$

45. $\left(-\dfrac{1}{2}+\dfrac{3}{4}i\right)-\left(\dfrac{3}{2}+\dfrac{1}{4}i\right)=-2+\dfrac{1}{2}i$

46. Perform the computation as you would with binomials, being sure to replace i^2 with -1 wherever it occurs. Thus, $(4+3i)(10-i)=40-4i+30i-3i^2=40-4i+30i+3=43+26i$.

47. $(-2-5i)(4-8i)=-8+16i-20i+40i^2=-48-4i$

48. $(x+yi)(x-yi)=x^2+y^2$. Thus, $(7+2i)(7-2i)=49+4=53$.

49. $(\sqrt{3}-i\sqrt{5})(\sqrt{3}+i\sqrt{5})=3+5=8$

50. $(x+yi)\left(\dfrac{x}{x^2+y^2}+\dfrac{-y}{x^2+y^2}i\right)=\dfrac{x^2}{x^2+y^2}+\dfrac{-xy}{x^2+y^2}i+\dfrac{xy}{x^2+y^2}i+\dfrac{-y^2}{x^2+y^2}i^2$

$$=\dfrac{x^2+y^2}{x^2+y^2}=1$$

51. $(i^4)^n=1$. Thus, $i^{302}=i^{300}i^2=(i^4)^{75}i^2=1\cdot-1=-1$.

52. Do the division by multiplying the numerator and denominator by the complex conjugate of the denominator. Thus, $\dfrac{1-2i}{3+4i}=\dfrac{(1-2i)}{(3+4i)}\cdot\dfrac{(3-4i)}{(3-4i)}=\dfrac{-5-10i}{9+16}=\dfrac{-5-10i}{25}=-\dfrac{1}{5}-\dfrac{2}{5}i$.

53. $\dfrac{4-2i}{2+3i}=\dfrac{(4-2i)}{(2+3i)}\cdot\dfrac{(2-3i)}{(2-3i)}=\dfrac{2-16i}{13}=\dfrac{2}{13}-\dfrac{16}{13}i$

54. $(5-3i)^{-1}=\dfrac{1}{(5-3i)}=\dfrac{1}{(5-3i)}\cdot\dfrac{(5+3i)}{(5+3i)}=\dfrac{5+3i}{34}=\dfrac{5}{34}+\dfrac{3}{34}i$

55. $(i)^{-1}=\dfrac{1}{i}\cdot\dfrac{-i}{-i}=\dfrac{-i}{-i^2}=\dfrac{-i}{1}=-i$

56. (A) $(4,8)$

(B) $(0,4)$

(C) $(-4,3)$

(D) $(0,0)$

(E) $(-5,-5)$

(F) $(0,-7)$

(G) $(3,-3)$

(H) $(8,0)$

57. distance $= \sqrt{(x_2 - x_1)^2 + (y_2 - y_1)^2}$

$$= \sqrt{(1-3)^2 + (7-4)^2} = \sqrt{(-2)^2 + (3)^2} = \sqrt{4+9} = \sqrt{13}$$

58. distance $= \sqrt{(x_2 - x_1)^2 + (y_2 - y_1)^2}$

$$= \sqrt{(6+2)^2 + (0+3)^2} = \sqrt{(8)^2 + (3)^2} = \sqrt{64+9} = \sqrt{73}$$

59. midpoint $= \left(\dfrac{x_1 + x_2}{2}, \dfrac{y_1 + y_2}{2} \right) = \left(\dfrac{7+5}{2}, \dfrac{10+6}{2} \right) = (6,8)$

60. midpoint $= \left(\dfrac{x_1 + x_2}{2}, \dfrac{y_1 + y_2}{2} \right) = \left(\dfrac{4-6}{2}, \dfrac{17+1}{2} \right) = (-1,9)$

61. (A) rise

(B) run

(C) negative

(D) positive

(E) zero

(F) undefined

62. slope $= m = \dfrac{y_2 - y_1}{x_2 - x_1} = \dfrac{-1-11}{4+2} = \dfrac{-12}{6} = -2$

63. slope $= m = \dfrac{y_2 - y_1}{x_2 - x_1} = \dfrac{0+4}{0+3} = \dfrac{4}{3}$

64. slope $= m = \dfrac{y_2 - y_1}{x_2 - x_1} = \dfrac{2-5}{-8-(-8)} = \dfrac{-3}{0} =$ undefined

65. slope $= m = \dfrac{y_2 - y_1}{x_2 - x_1} = \dfrac{6-6}{-10-4} = \dfrac{0}{-14} = 0$

Chapter 2: Precalculus Function Skills

66. Recall that a function is a set of ordered pairs in which each first component is paired with *exactly one* second component. That is, in a function no two ordered pairs have the same first component but different second components.

(A) True, because $\{(-4,5),(-1,5),(0,5),(5,5)\}$ is a set of ordered pairs in which each first component is paired with exactly one second component, so it is a function.

(B) False, because, for instance, if x is 8, then $y^2 = 16$ and $y = \pm 4$, so the set $\{(x,y) \mid y^2 = 2x\}$ contains the ordered pairs $(8,4)$ and $(8,-4)$; thus, it is not a function.

(C) True. If $(4, a)$ and $(4, b)$ are elements of a function, then $a = b$ because the first component of a function cannot be paired with two different second components.

(D) False. A function is a set of ordered pairs. The domain of a function is the set composed of the first elements of the function; the domain is not a set of ordered pairs.

(E) True. In the function $f = \{(x, y) \mid y = 5x + 3\}$, $y = f(x)$ is the image of x under f.

67. $y = f\left(\dfrac{3}{4}\right) = 28\left(\dfrac{3}{4}\right) - 10 = 21 - 10 = 11$

68. $y = f(-5) = (-5)^2 + 1 = 25 + 1 = 26$

69. $y = f(-1) = 4(-1)^5 + 2(-1)^4 - 3(-1)^3 - 5(-1)^2 + (-1) + 5 = 0$

70. $y = f\left(-\dfrac{\pi}{4}\right) = \dfrac{180\,|x|}{\pi} = \dfrac{180\left|-\dfrac{\pi}{4}\right|}{\pi} = \dfrac{180 \cdot \dfrac{\pi}{4}}{\pi} = 45$

71. $y = f(3) = \dfrac{4(3) - 5}{(3)^2 + 1} = \dfrac{7}{10}$

72. $f(5x) = \dfrac{2(5x) - 3}{(5x) + 1} = \dfrac{10x - 3}{5x + 1}$

73. $f(x^2 + 1) = \dfrac{2(x^2 + 1) - 3}{(x^2 + 1) - 1} = \dfrac{2x^2 - 1}{x^2}$

74. The domain of a function $f = \{(x, y) \mid y = f(x)\}$ is the set of all possible x values of f and the range is the set of all possible y values of f. Because the square root of a negative number is not a real number, to find the domain of f solve $3x - 15 \geq 0$ for x. Thus, $D_f = \{x \mid x \geq 5\}$. Because the square root always returns a nonnegative number, $R_f = \{y \mid y \geq 0\}$.

75. $D_f = \{4, 5, 5.2, 10\}$; $R_f = \left\{-1, -\dfrac{3}{4}, \sqrt{3}, 12\right\}$

76. $g(x) = \dfrac{5}{x + 4}$ is undefined when $x = -4$ because division by zero would occur. Thus, $D_g = \{x \mid x \neq -4\}$. Solving $y = g(x) = \dfrac{5}{x + 4}$ for x yields $x = \dfrac{5 - 4y}{y}$, which is undefined when $y = 0$. Thus, $R_g = \{y \mid y \neq 0\}$.

77. The absolute value is defined for all real numbers and always returns a nonnegative number, so $D_h = R$ and $R_h = \{y \mid y \geq 0\}$.

78. Because the square root of a negative number is not a real number, to find the domain of f determine when $x^2 - 9$, which equals the product $(x + 3)(x - 3)$, is nonnegative. The product is zero at $x = -3$ or $x = 3$. It is positive when both factors have the same sign, which occurs to the left of -3 and to right of 3. Therefore, $D_f = \{x \mid x \le -3 \text{ or } x \ge 3\}$. Because the square root always returns a nonnegative number, $R_f = \{y \mid y \ge 0\}$.

79. $(f + g)(x) = f(x) + g(x) = \dfrac{1}{x} + x^4, \ x \ne 0$

80. $(f - g)(x) = f(x) - g(x) = \dfrac{1}{x} - x^4, \ x \ne 0$

81. $(fg)(x) = f(x)g(x) = \left(\dfrac{1}{x}\right)(x^4) = x^3, \ x \ne 0$

82. $\left(\dfrac{f}{g}\right)(x) = \dfrac{f(x)}{g(x)} = \dfrac{\frac{1}{x}}{x^4} = \dfrac{1}{x^5}, \ x \ne 0$

83. $(f + g)(x) = (x^2 + 3) + (\sqrt{x} - 3) = x^2 + \sqrt{x}, \ x \ge 0$.
Thus, $(f + g)(4) = 4^2 + \sqrt{4} = 16 + 2 = 18$.

84. $(f - g)(x) = (x^2 + 3) - (\sqrt{x} - 3) = x^2 - \sqrt{x} + 6, \ x \ge 0$.
Thus, $(f - g)(6) = 6^2 - \sqrt{6} + 6 = 36 - \sqrt{6} + 6 = 42 - \sqrt{6}$.

85. $(fg)(-1)$ is undefined because -1 is not in the domain of fg.

86. $\left(\dfrac{f}{g}\right)(x) = \dfrac{x^2 + 3}{\sqrt{x} - 3}, x \ge 0, x \ne 9$. Notice that in addition to being nonnegative,

the domain of $\dfrac{f}{g}$ excludes the value 9 because it must exclude values of x for which

$\sqrt{x} - 3 = 0$. Thus, $\left(\dfrac{f}{g}\right)(9)$ is undefined because 9 is not in the domain of $\dfrac{f}{g}$.

87. The composition of two functions f and g is the function $f \circ g$ defined by $(f \circ g)(x) = f(g(x))$, provided that $g(x) \in D_f$. Thus, the function g takes x into $g(x)$ and the function f takes $g(x)$ into $f(g(x))$. Obviously, any value of x for which $g(x)$ is not in the domain of f cannot be in the domain of $(f \circ g)(x)$.

Looking at the ordered pairs of f and g, you can see that the function g takes -4 and 3 into 2 and -4, respectively; and the function f takes 2 and -4 into 5 and -3, respectively. Thus, (a) $f \circ g = \{(-4, 5), (3, -3)\}$; (b) $(f \circ g)(-4) = 5$.

88. From the ordered pairs of f and g, you can see that the function f takes $-5, -4, -1$, and 6 into $1, -3, 3$, and -4, respectively; and the function g takes $1, -3, 3$, and -4 into $9, -3$, -4, and 2, respectively. Thus, (a) $g \circ f = \{(-5, 9), (-4, -3), (-1, -4), (6, 2)\}$; (b) $(g \circ f)(6) = 2$.

89. Proceeding as in question 87, (a) $f \circ g = \{(0,-8),(1,8),(2,1)\}$; (b) $(f \circ g)(1) = 8$.

90. Proceeding as in question 88, (a) $g \circ f = \{(-1,8),(0,2)\}$; (b) $(g \circ f)(2)$ is undefined because 2 is not in the domain of $g \circ f$.

91. (a) $(f \circ g)(x) = f(g(x)) = \sqrt{g(x) + 5} = \sqrt{(3x^2) + 5} = \sqrt{3x^2 + 5}$; (b) domain $= R$

92. (a) $(f \circ g)(x) = f(g(x)) = |g(x)| = |9x - 4|$; (b) domain $= R$

93. (a) $(f \circ g)(x) = f(g(x)) = \sqrt{g(x)} = \sqrt{16x^2} = 4|x|$. *Note:* $\sqrt{x^2} = |x|$, not x or $\pm x$; (b) domain $= R$

94. (a) $(f \circ g)(x) = f(g(x)) = \dfrac{1}{g(x) + 1} = \dfrac{1}{5x + 1}$; (b) domain $= \left\{ x \,\middle|\, x \neq -\dfrac{1}{5} \right\}$

95. (a) $(f \circ g)(x) = f(g(x)) = \dfrac{1 - g(x)}{12} = \dfrac{1 - (1 - 12x)}{12} = \dfrac{12x}{12} = x$; (b) domain $= R$

96. $(f \circ g)(x) = \sqrt{g(x) - 4} = \sqrt{(x^2 + 4) - 4} = \sqrt{x^2} = |x|$. Thus, $(f \circ g)(9) = |9| = 9$

97. $(f \circ g)(0) = |0| = 0$

98. $(f \circ f)(x) = \sqrt{f(x) - 4} = \sqrt{\sqrt{x - 4} - 4}$, $x \geq 20$. *Note:* For f, $x - 4 \geq 0$; that is, $x \geq 4$. And for $f \circ f$, $\sqrt{x - 4} - 4 \geq 0$, which solves to give $x \geq 20$. So, $f \circ f$ has the domain $\{x \mid x \geq 20\}$. Thus, $(f \circ f)(40) = \sqrt{\sqrt{40 - 4} - 4} = \sqrt{\sqrt{36} - 4} = \sqrt{6 - 4} = \sqrt{2}$.

99. $(f \circ f)(10)$ is undefined because 10 is not in the domain of $f \circ f$.

100. $(g \circ f)(x) = (f(x))^2 + 4 = (\sqrt{x - 4})^2 + 4 = x - 4 + 4 = x$, $x \geq 4$. Thus, $(g \circ f)(5) = 5$.

101. A function is one-to-one if each first component is paired with *exactly one* second component, *and* each second component is paired with *exactly one* first component. Thus, f is one-to-one.

102. g is one-to-one.

103. h is not one-to-one because $\left(2, \dfrac{3}{4} \right)$ and $(4, 0.75)$ have the same second component.

104. f is one-to-one because no two different x values yield the same y value.

105. g is not one-to-one because, for instance, both 1 and -1 yield the same y value of 5.

106. If the function f is one-to-one, the inverse of f is the function f^{-1} (read "f inverse") whose ordered pairs are obtained from f by interchanging the first and second components of each of the ordered pairs in f. Accordingly, the domain of f^{-1} is the range of f, and the range of f^{-1} is the domain of f. Thus, $f^{-1} = \{(1,0),(5,1),(9,2),(13,3)\}$. *Note:* The notation f^{-1} does not indicate a reciprocal. The raised $^{-1}$ is not an exponent. Do NOT write $f^{-1} = \dfrac{1}{f}$.

107. Method 1. Set $(f \circ f^{-1})(x) = x$ and solve for $f^{-1}(x)$.

$$(f \circ f^{-1})(x) = x$$
$$f(f^{-1}(x)) = x$$
$$3f^{-1}(x) + 1 = x$$
$$3f^{-1}(x) = x - 1$$
$$f^{-1}(x) = \frac{x-1}{3}$$

Method 2. Interchange x and y in $y = 3x + 1$, and then solve for y.

$$x = 3y + 1$$
$$3y + 1 = x$$
$$3y = x - 1$$
$$y = f^{-1}(x) = \frac{x-1}{3}$$

108. Using Method 2 from question 107, interchange x and y in $y = (x+5)^3$, and then solve for y.

$$x = (y+5)^3$$
$$(y+5)^3 = x$$
$$y + 5 = \sqrt[3]{x}$$
$$y = g^{-1}(x) = \sqrt[3]{x} - 5$$

109. Using Method 1 from question 107:

$$(h \circ h^{-1})(x) = x$$
$$h(h^{-1}(x)) = x$$
$$7h^{-1}(x) = x$$
$$h^{-1}(x) = \frac{x}{7}$$

110. Using Method 1 from question 107:

$$(g \circ g^{-1})(x) = x$$
$$g(g^{-1}(x)) = x$$
$$\frac{g^{-1}(x) + 2}{g^{-1}(x) - 1} = x$$
$$x(g^{-1}(x) - 1) = g^{-1}(x) + 2$$
$$xg^{-1}(x) - x = g^{-1}(x) + 2$$
$$xg^{-1}(x) - g^{-1}(x) = x + 2$$
$$g^{-1}(x)(x - 1) = x + 2$$
$$g^{-1}(x) = \frac{x + 2}{x - 1}$$

111. If a function is not one-to-one, you might restrict the domain so that the function is one-to-one in the restricted domain as in this problem.

Given $y = x^2$ and $x \geq 0$, you have $\sqrt{y} = x$.

Using Method 2 from question 107, interchange x and y in $\sqrt{y} = x$, and then solve for y.

$$\sqrt{x} = y$$
$$y = f^{-1}(x) = \sqrt{x}$$

112. $f^{-1} = \{(1,0), (5,1), (9,2), (13,3)\}$; thus, $f^{-1}(9) = 2$.

113. $f^{-1}(x) = \frac{9}{5}x + 32$; thus, $f^{-1}(0) = \frac{9}{5}(0) + 32 = 32$.

114. $f^{-1}(x) = \frac{4x - 20}{3}$; thus, $f^{-1}(0.5) = \frac{4(0.5) - 20}{3} = \frac{2 - 20}{3} = \frac{-18}{3} = -6$.

115. $f^{-1}(x) = \sqrt[3]{2x - 6}$; thus, $f^{-1}(-29) = \sqrt[3]{2(-29) - 6} = \sqrt[3]{-58 - 6} = \sqrt[3]{-64} = -4$.

116. A function f is even if $f(-x) = f(x)$ for every $x \in D_f$; f is odd if $f(-x) = -f(x)$ for every $x \in D_f$. *Note*: The graphs of even functions are symmetric about the y-axis, and those of odd functions are symmetric about the origin. The function f defined by $f(x) = x^3$ is odd because $f(-x) = (-x)^3 = -x^3 = -f(x)$.

117. The function g defined by $g(x) = |x + 3|$ is neither even nor odd because $g(-x) = |-x + 3|$, which does not simplify to be either $g(x)$ or $-g(x)$.

118. The function h defined by $h(x) = 36x^2 - 10$ is even because $h(-x) = 36(-x)^2 - 10 = 36x^2 - 10 = h(x)$.

119. The function g defined by $g(x) = 2x^2 - x + 1$ is neither even nor odd because $g(-x) = 2(-x)^2 - (-x) + 1 = 2x^2 + x + 1$, which does not simplify to be either $g(x)$ or $-g(x)$.

120. The function f defined by $f(x) = 3x^4 - 10x^2 + 8$ is even because $f(-x) = 3(-x)^4 - 10(-x)^2 + 8 = 3x^4 - 10x^2 + 8 = f(x)$.

Chapter 3: Graphs of Functions

121. domain $= (-\infty, \infty)$, range $= \{y \mid y \geq -4\}$

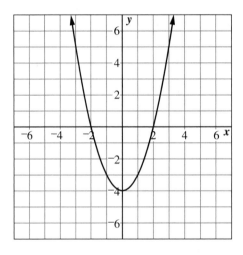

122. domain $= \{x \mid x \geq 2\}$, range $= \{y \mid y \geq 0\}$

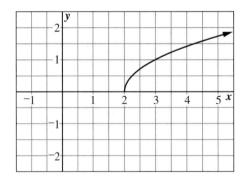

123. domain $= (-\infty, \infty)$, range $= (-\infty, \infty)$

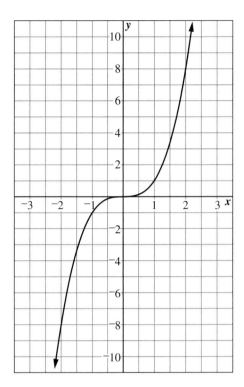

124. domain $= (-\infty, \infty)$, range $= \{y \mid 0 < y \le 2\}$

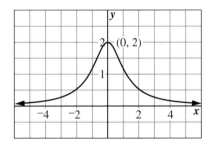

125. The vertical line test specifies that a graph represents a function if and only if no vertical line intersects the graph in more than one point. The graph shown passes the vertical line test, so it is a function.

126. Proceeding as in question 125, the graph is not a function. It does not pass the vertical line test.

127. To determine the x-intercept(s) for a function f, set $f(x) = 0$ and then solve for x. Similarly, provided that 0 is in the domain of f, to determine the y-intercept let $x = 0$, and then solve for y. For $f(x) = 6x - 1$, solving $6x - 1 = 0$ yields $x = \dfrac{1}{6}$ as the x-intercept. Solving $y = 6(0) - 1$ yields $y = -1$ as the y-intercept.

128. Solving $6x^2 + 5x - 4 = (2x - 1)(3x + 4) = 0$ yields $x = \dfrac{1}{2}$, and $-\dfrac{4}{3}$ as x-intercepts. Solving $y = 6 \cdot 0^2 + 5 \cdot 0 - 4 = -4$ yields $y = -4$ as the y-intercept.

129. Solving $(x + 1)(x + 3)(x - 2)(x - 4) = 0$ yields $x = -3, -1, 2,$ and 4 as x-intercepts. Solving $y = (0 + 1)(0 + 3)(0 - 2)(0 - 4)$ yields $y = 24$ as the y-intercept.

130. Solving $x^2 + 4 = 0$ yields no solution because $x^2 + 4 \geq 4$; thus, there is no x-intercept. Solving $y = 0^2 + 4$ yields $y = 4$ as the y-intercept.

131. Moving from left to right on the graph of f shown, the y values decrease until $x = 1$ and increase thereafter. Thus f is decreasing on $(-\infty, 1]$ and increasing on $[1, \infty)$.

132. f is constant and is neither increasing nor decreasing.

133. Moving from left to right, the y values increase until $x = -0.6$, decrease until $x = 0$, and increase thereafter. Thus, f is increasing on $(-\infty, -0.6]$, decreasing on $[-0.6, 0]$, and increasing on $[0, \infty)$.

134. The lowest point of the graph is at $(0, -4)$. Therefore, -4 is an absolute minimum of f.

135. In a local interval around $x = -0.6$, the y value of 0 is locally the greatest, so f has a relative maximum of 0. Similarly, in a local interval around 0, the y value of -1 is locally the least, so f has a relative minimum of -1.

136. f has no relative or absolute extrema.

137. The highest point of the graph is at $(0,2)$. Therefore, 2 is an absolute maximum of f.

138. $g(x) = \sqrt{x} + 7$

139. $g(x) = \sqrt{x-7}$

140. $g(x) = |x| - \dfrac{2}{5}$

141. $g(x) = \left| x + \dfrac{2}{5} \right|$

142. $g(x) = \dfrac{1}{x-4}$

143. A reflection about the x-axis of the graph defined by $y = f(x)$ is given by $y = -f(x)$. Thus, $g(x) = -f(x) = -|x+3|$.

144. A reflection about the y-axis of the graph defined by $y = f(x)$ is given by $y = f(-x)$. Thus, $g(x) = f(-x) = |-x+3|$.

145. $g(x) = -f(x) = -5x^2 - 3x + 1$

146. $g(x) = f(-x) = 5x^2 - 3x - 1$

147. $g(x) = \dfrac{0.2}{x^2} = 0.2\left(\dfrac{1}{x^2}\right) = 0.2 f(x)$ is (b) a vertical compression of $f(x) = \dfrac{1}{x^2}$.

148. $g(x) = \dfrac{2}{x^2} = 2\left(\dfrac{1}{x^2}\right) = 2 f(x)$ is (a) a vertical stretch of $f(x) = \dfrac{1}{x^2}$.

149. $g(x) = 15x^2 = 15(x^2) = 15 f(x)$ is (a) a vertical stretch of $f(x) = x^2$.

150. $g(x) = \sqrt{0.2x} = f(0.2x)$ is (c) a horizontal stretch of $f(x) = \sqrt{x}$.

151. $g(x) = \sqrt{5x} = f(5x)$ is (d) a horizontal compression of $f(x) = \sqrt{x}$.

152. $g(x) = f(2x) = \dfrac{1}{2x} - 3$

153. $g(x) = 3f(x) = 3\sqrt{x+1}$

154. $g(x) = f\left(\dfrac{1}{2}x\right) = \left(\dfrac{1}{2}x\right)^3$

155. $g(x) = 100 f(x) = 100\sqrt{x}$

156. If (x, y) is on the graph of $y = f(x)$, then (x, ay) is on the graph of $g(x) = af(x)$. Thus, given that $\left(2, \dfrac{1}{4}\right)$ is on the graph of $f(x) = \dfrac{1}{x^2}$, then $\left(2, \ 12 \cdot \dfrac{1}{4}\right) = (2, 3)$ is on the graph of $g(x) = 12\left(\dfrac{1}{x^2}\right)$.

157. Given that $\left(6, \dfrac{1}{6}\right)$ is on the graph of $f(x) = \dfrac{1}{x}$, then $\left(6, \ 2 \cdot \dfrac{1}{6}\right) = \left(6, \dfrac{1}{3}\right)$ is on the graph of $g(x) = \dfrac{2}{x}$.

158. If (x, y) is on the graph defined by $y = f(x)$, then $\left(\dfrac{x}{b}, y\right)$ is on the graph defined by $g(x) = f(bx)$. Thus, given that $(-4, 4)$ is on the graph of $f(x) = |x|$, then $\left(\dfrac{-4}{0.25}, 4\right) = (-16, 4)$ is on the graph of $g(x) = |0.25x|$.

159. Given that $(-5, 25)$ is on the graph of $f(x) = x^2$, then $(-5, \ 10 \cdot 25) = (-5, 250)$ is on the graph of $g(x) = 10x^2$.

160. Given that $(36, 6)$ is on the graph of $f(x) = \sqrt{x}$, then $\left(\dfrac{36}{3}, 6\right) = (12, 6)$ is on the graph of $g(x) = \sqrt{3x}$.

161. $g(x) = 10\left[\dfrac{1}{(-x-5)^3}\right] + 3$

162. $g(x) = -0.25\,|x + 2.5| + 6.75$

163. $g(x) = 5[2(x-7)]^3 - 8$

164. $g(x) = 2(x+5)^2 + 3(x+5) + 9$

165. $g(x) = 6\sqrt{\dfrac{1}{3}(x+3)} - 5$

Chapter 4: Linear and Quadratic Functions, Equations, and Inequalities

166. (a) x-intercept $= -\dfrac{b}{m} = -\dfrac{5}{3}$, y-intercept $= b = 5$; (b) $x = -\dfrac{5}{3}$ is the zero of f.

167. (a) x-intercept $= -\dfrac{b}{m} = -\dfrac{28}{3}$, y-intercept $= b = -7$; (b) $x = -\dfrac{28}{3}$ is the zero of f.

168. (a) Linear functions of the form $f(x) = b$ are constant functions. When $b \neq 0$, the graph of a constant function is a horizontal line that is $|b|$ units above or below the x-axis with slope 0 and y-intercept b. It has no x-intercepts and no zeros. Thus, for $f(x) = 10$, x-intercept = none, y-intercept = 10; (b) no zeros.

169. (a) In a constant function when $b = 0$, the graph is coincident with the x-axis. Thus, $f(x) = 0$ has y-intercept, 0, and every real number is an x-intercept; (b) all real numbers are zeros of f.

170. (a) x-intercept $= -\dfrac{5}{2}$, y-intercept $= -5$; (b) $x = -\dfrac{5}{2}$ is the zero of f.

171. Substituting into the point-slope form of a linear equation,

$$y - y_1 = m(x - x_1)$$
$$y - (-1) = \frac{3}{4}(x - (-8))$$
$$y + 1 = \frac{3}{4}x + 6$$
$$y = \frac{3}{4}x + 5$$

172. $y - y_1 = m(x - x_1)$
$$y - 2 = -4(x - (-5))$$
$$y - 2 = -4x - 20$$
$$y = -4x - 18$$

173. slope $= m = \dfrac{y_2 - y_1}{x_2 - x_1} = \dfrac{-1-11}{4+2} = \dfrac{-12}{6} = -2$. Now, using the point $(4,-1)$ and substituting into the point-slope form of a linear equation,

$$y - y_1 = m(x - x_1)$$
$$y - (-1) = -2(x - 4)$$
$$y + 1 = -2x + 8$$
$$y = -2x + 7$$

Note: Instead of using $(4,-1)$, you could use the other point $(-2,11)$ and obtain the same result in this problem.

174. slope $= m = \dfrac{y_2 - y_1}{x_2 - x_1} = \dfrac{6-6}{-10-4} = \dfrac{0}{-14} = 0$. Now, using the point $(4, 6)$ and substituting into the point-slope form of linear equation,

$$y - y_1 = m(x - x_1)$$
$$y - 6 = 0(x - 4)$$
$$y = 6$$

175. $x + 3(x - 2) = 2x - 4$

$x + 3x - 6 = 2x - 4$	Remove parentheses.
$4x - 6 = 2x - 4$	Simplify.
$4x - 6 - 2x = 2x - 4 - 2x$	Subtract $2x$ from both sides.
$2x - 6 = -4$	Simplify.
$2x - 6 + 6 = -4 + 6$	Add 6 to both sides.
$2x = 2$	Simplify.
$\dfrac{2x}{2} = \dfrac{2}{2}$	Divide both sides by 2.
$x = 1$	Simplify.

176.

$$\frac{z}{5} - 3 = \frac{3}{10} - z$$

$$\frac{10}{1}\left(\frac{z}{5} - 3\right) = \frac{10}{1}\left(\frac{3}{10} - z\right)$$ Remove fractions by multiplying both sides by 10.

$$2z - 30 = 3 - 10z$$

$$2z - 30 + 10z = 3 - 10z + 10z$$ Add $10z$ to both sides.

$$12z - 30 = 3$$ Simplify.

$$12z - 30 + 30 = 3 + 30$$ Add 30 to both sides.

$$12z = 33$$ Simplify.

$$\frac{12z}{12} = \frac{33}{12}$$ Divide both sides by 12.

$$z = \frac{11}{4}$$ Simplify.

177.

$$\frac{x+3}{5} = \frac{x-1}{2}$$

$$\frac{10}{1} \cdot \frac{x+3}{5} = \frac{10}{1} \cdot \frac{x-1}{2}$$ Multiple both sides by 10.

$$2(x+3) = 5(x-1)$$ Simplify. *Note*: Shortcut is to "cross multiply."

$$2x + 6 = 5x - 5$$ Remove parentheses.

$$2x + 6 - 5x = 5x - 5 - 5x$$ Subtract $5x$ from both sides.

$$-3x + 6 = -5$$ Simplify.

$$-3x + 6 - 6 = -5 - 6$$ Subtract 6 from both sides.

$$-3x = -11$$ Simplify.

$$\frac{-3x}{-3} = \frac{-11}{-3}$$ Divide both sides by -3.

$$x = \frac{11}{3}$$ Simplify.

178.

$$Ax + By = C$$

$$Ax + By - Ax = C - Ax$$ Isolate the variable term by subtracting Ax from both sides.

$$By = C - Ax$$ Simplify.

$$\frac{By}{B} = \frac{C - Ax}{B}$$ Divide both sides by B.

$$y = \frac{C}{B} - \frac{Ax}{B} = -\frac{A}{B}x + \frac{C}{B}$$ Write the answer in slope-intercept form.

179.

$$x = \frac{y+2}{y-1}$$

$$\left(\frac{y-1}{1}\right) \cdot x = \left(\frac{y-1}{1}\right)\left(\frac{y+2}{y-1}\right)$$ Multiply both sides by $\frac{y-1}{1}$.

$$(y-1) \cdot x = y + 2$$ Simplify.

$$yx - x = y + 2$$ Remove parentheses.

$$yx - x - y = y + 2 - y$$ Subtract y from both sides.

$$yx - x - y = 2$$ Simplify.

$$yx - x - y + x = 2 + x$$ Add x to both sides.

$$yx - y = 2 + x$$ Simplify.

$$y(x-1) = x + 2$$ Factor the side containing y.

$$\frac{y(x-1)}{x-1} = \frac{x+2}{x-1}$$ Divide both sides by $x - 1$.

$$y = \frac{x+2}{x-1}$$ Simplify.

180.

$$3x + 2 > 6x - 4$$

$$3x + 2 - 6x > 6x - 4 - 6x$$

$$-3x + 2 > -4$$

$$-3x + 2 - 2 > -4 - 2$$

$$-3x > -6$$

$$\frac{-3x}{-3} < \frac{-6}{-3}$$ Note the reversal of the inequality.

$$x < 2$$

181.
$$3x - 2 \le 7 - 2x$$
$$3x - 2 + 2x \le 7 - 2x + 2x$$
$$x - 2 \le 7$$
$$x - 2 + 2 \le 7 + 2$$
$$x \le 9$$

182.
$$\frac{x+3}{5} \ge \frac{x-1}{2}$$
$$\frac{10}{1} \cdot \frac{x+3}{5} \ge \frac{10}{1} \cdot \frac{x-1}{2}$$
$$2(x+3) \ge 5(x-1)$$
$$2x + 6 \ge 5x - 5$$
$$2x + 6 - 5x \ge 5x - 5 - 5x$$
$$-3x + 6 \ge -5$$
$$-3x + 6 - 6 \ge -5 - 6$$
$$-3x \ge -11$$
$$\frac{-3x}{-3} \le \frac{-11}{-3} \qquad \text{Note the reversal of the inequality.}$$
$$x \le \frac{11}{3}$$

183. (a) y-intercept = 30; (b) $b^2 - 4ac = (-10)^2 - 4(1)(30) = 100 - 120 = -20 < 0$, so there are no real zeros and, thus, no x-intercepts.

184. (a) y-intercept = -5; (b) $b^2 - 4ac = (2)^2 - 4(2)(-5) = 44 > 0$, so there are two real zeros and, thus, two x-intercepts.

185. (a) y-intercept = 16; (b) $b^2 - 4ac = (-8)^2 - 4(1)(16) = 0$, so there is one real zero and, thus, one x-intercept.

186. (a) The vertex for the quadratic form $f(x) = ax^2 + bx + c$ is $\left(-\frac{b}{2a}, f\left(-\frac{b}{2a} \right) \right)$. Thus, for $f(x) = x^2 - 10x + 30, -\frac{b}{2a} = -\frac{-10}{2(1)} = 5$ and $f(5) = (5)^2 - 10(5) + 30 = 5$, so the vertex is $(5,5)$; (b) The parabolic graph of a quadratic function is symmetric about a vertical line through its vertex. This line, with equation $x = -\frac{b}{2a}$, is the axis of symmetry. Thus, $x = 5$ is the equation for the axis of symmetry for $f(x) = x^2 - 10x + 30$.

187. (a) $-\dfrac{2}{2(2)} = -\dfrac{1}{2}$ and $g\left(-\dfrac{1}{2}\right) = 2\left(-\dfrac{1}{2}\right)^2 + 2\left(-\dfrac{1}{2}\right) - 5 = -\dfrac{11}{2}$, so the vertex is $\left(-\dfrac{1}{2}, -\dfrac{11}{2}\right)$; (b) $x = -\dfrac{1}{2}$ is the equation for the axis of symmetry.

188. (a) $-\dfrac{b}{2a} = -\dfrac{-8}{2(1)} = 4$ and $h(4) = (4)^2 - 8(4) + 16 = 0$, so the vertex is $(4,0)$; (b) $x = 4$ is the equation for the axis of symmetry.

189. (a) The vertex form for $f(x) = ax^2 + bx + c$ is $f(x) = a(x - h)^2 + k$. Convert to vertex form by completing the square:

$f(x) = 2(x^2 - 6x + \quad) + 17$

$f(x) = 2(x^2 - 6x + 9) + 17 - 18$ *Note:* Add and subtract 2 times the square of $\frac{1}{2}$ of x's coefficient.

$f(x) = 2(x - 3)^2 - 1$

(b) The vertex of the parabolic graph of $f(x) = a(x - h)^2 + k$ is (h,k). Thus, the vertex of $f(x) = 2(x - 3)^2 - 1$ is $(3,-1)$.

190. (a) $h(x) = x^2 - 8x + 16$

$\quad\quad h(x) = (x - 4)^2$

(b) $(4,0)$

191. (a) $g(x) = -x^2 + 10x + 3$

$\quad\quad g(x) = -(x^2 - 10x + \quad) + 3$

$\quad\quad g(x) = -(x^2 - 10x + 25) + 3 + 25$

$\quad\quad g(x) = -(x - 5)^2 + 28$

(b) $(5,28)$

192. (A) upward, downward
(B) maximum
(C) minimum

193. (a) upward; (b) $(6,-31)$; (c) $x = 6$; (d) minimum $y = -31$; (e) domain $= R$, range $= \{y \mid y \geq -31\}$; (f) decreasing on $(-\infty, \ 6]$ and increasing on $[6, \ \infty)$

194. (a) upward; (b) $(0,4)$; (c) $x = 0$; (d) minimum $y = 4$; (e) domain $= R$, range $= \{y \mid y \geq 4\}$; (f) decreasing on $(-\infty, \ 0]$ and increasing on $[0, \ \infty)$

195. (a) downward; (b) $(-1,-7)$; (c) $x = -1$; (d) maximum $y = -7$; (e) domain $= R$, range $= \{y \mid y \leq -7\}$; (f) increasing on $(-\infty, \ -1]$ and decreasing on $[-1, \ \infty)$

196. (a) downward; (b) $(5,28)$; (c) $x = 5$; (d) maximum $y = 28$; (e) domain $= R$, range $= \{y \mid y \leq 28\}$; (f) increasing on $(-\infty, \ 5]$ and decreasing on $[5, \ \infty)$

197.

$$x^2 - x - 6 = 0$$

$$(x - 3)(x + 2) = 0 \qquad \text{Factor the quadratic expression.}$$

$$x - 3 = 0 \text{ or } x + 2 = 0 \qquad \text{Set each factor equal to zero.}$$

$$x = 3 \text{ or } -2 \qquad \text{Solve for } x.$$

198. $x^2 + 6x - 1 = -5$

$$x^2 + 6x = -4 \qquad \text{Isolate the variable terms.}$$

$$x^2 + 6x + 9 = -4 + 9 \qquad \text{Add the square of } \tfrac{1}{2} \text{ of } x\text{'s coefficient to both sides.}$$

$$(x + 3)^2 = 5 \qquad \text{Factor the perfect square on the left side, and simplify the right side.}$$

$$x + 3 = \pm\sqrt{5} \qquad \text{Take the square root of both sides.}$$

$$x = -3 \pm \sqrt{5} \qquad \text{Solve for } x.$$

$$x = -3 + \sqrt{5} \text{ or } -3 - \sqrt{5}$$

199. $3x^2 - 5x + 1 = 0$

$$a = 3, \ b = -5, \text{ and } c = 1 \qquad \text{Identify the coefficients.}$$

Note: Keep $-$ symbol with number that follows it.

$$x = \frac{-b \pm \sqrt{b^2 - 4ac}}{2a} = \frac{-(-5) \pm \sqrt{25 - 4(3)(1)}}{6} \qquad \text{Substitute into the quadratic formula.}$$

$$x = \frac{5 \pm \sqrt{13}}{6} \qquad \text{Evaluate and simplify.}$$

$$x = \frac{5 + \sqrt{13}}{6} \text{ or } \frac{5 - \sqrt{13}}{6}$$

200. $x^2 - 3x + 2 = 0$

$(x - 2)(x - 1) = 0$

$x - 2 = 0$ or $x - 1 = 0$

$x = 2$ or 1

201. $9x^2 + 18x - 17 = 0$

$a = 9$, $b = 18$, and $c = -17$

$x = \dfrac{-b \pm \sqrt{b^2 - 4ac}}{2a} = \dfrac{-18 \pm \sqrt{18^2 - 4(9)(-17)}}{18}$

$x = \dfrac{-18 \pm \sqrt{936}}{18}$

$x = \dfrac{-18 \pm \sqrt{36(26)}}{18}$

$x = \dfrac{-18 \pm 6\sqrt{26}}{18}$

$x = \dfrac{-3 \pm \sqrt{26}}{3}$

$x = \dfrac{-3 + \sqrt{26}}{3}$ or $\dfrac{-3 - \sqrt{26}}{3}$

202. $6x^2 - 12x + 7 = 0$

$a = 6$, $b = -12$, and $c = 7$

$x = \dfrac{-b \pm \sqrt{b^2 - 4ac}}{2a} = \dfrac{12 \pm \sqrt{144 - 4(6)(7)}}{12}$

$x = \dfrac{12 \pm \sqrt{-24}}{12} = \dfrac{12 \pm 2i\sqrt{6}}{12} = \dfrac{6 \pm i\sqrt{6}}{6}$

$x = 1 + \dfrac{\sqrt{6}}{6}i$ or $1 - \dfrac{\sqrt{6}}{6}i$

203. $x^2 - x - 12 < 0$

$(x - 4)(x + 3) < 0$

real roots: $x = -3,\ 4$

solution set: $(-3, 4)$

204. $-x^2 + x + 12 < 0$

$\quad\quad x^2 - x - 12 > 0$ $\quad\quad\quad\quad$ Multiply both sides by −1 to make the coefficient of x^2 positive.

$\quad\quad (x - 4)(x + 3) > 0$

$\quad\quad$ real roots: $x = -3, 4$

$\quad\quad$ solution set: $(-\infty, -3) \cup (4, \infty)$

205. $x^2 - 10x + 30 > 0$

$\quad\quad$ real roots: none, because $b^2 - 4ac = (-10)^2 - 4(1)(30) = 100 - 120 = -20 < 0$

$\quad\quad$ solution set: $(-\infty, \infty)$

206. $x^2 - 10x + 25 \geq 0$

$\quad\quad$ real root: $x = 5$

$\quad\quad$ solution set: $(-\infty, \infty)$

207. $x^2 - 5 \geq 0$

$\quad\quad (x - \sqrt{5})(x + \sqrt{5}) \geq 0$

$\quad\quad$ real roots: $x = -\sqrt{5}, \sqrt{5}$

$\quad\quad$ solution set: $(-\infty, -\sqrt{5}] \cup [\sqrt{5}, \infty)$

208. $\dfrac{f(x_2) - f(x_1)}{x_2 - x_1} = \dfrac{f(3) - f(-3)}{3 - (-3)} = \dfrac{(-2(3) - 5) - (-2(-3) - 5)}{3 - (-3)} = \dfrac{-12}{6} = -2$

209. $\dfrac{f(x_2) - f(x_1)}{x_2 - x_1} = \dfrac{f(8) - f(-4)}{8 - (-4)} = \dfrac{\left(\frac{1}{2}(8) + 3\right) - \left(\frac{1}{2}(-4) + 3\right)}{8 - (-4)} = \dfrac{6}{12} = \dfrac{1}{2}$

210. $\dfrac{f(x_2) - f(x_1)}{x_2 - x_1} = \dfrac{f(5) - f(-5)}{5 - (-5)} = \dfrac{(10) - (10)}{5 - (-5)} = \dfrac{0}{10} = 0$

211. $\dfrac{f(x_2) - f(x_1)}{x_2 - x_1} = \dfrac{f(5) - f(-3)}{5 - (-3)} = \dfrac{k(5) - k(-3)}{5 - (-3)} = \dfrac{8k}{8} = k$

Note: $f(x) = kx$ is a directly proportional function, where nonzero $k \in R$ is the constant of proportionality.

212. $\dfrac{f(x_2)-f(x_1)}{x_2-x_1}=\dfrac{f(-4)-f(-8)}{-4-(-8)}=\dfrac{((-4)^2+8(-4)-5)-((-8)^2+8(-8)-5)}{-4-(-8)}$

$$=\dfrac{-21+5}{4}=\dfrac{-16}{4}=-4$$

213. $\dfrac{f(x_2)-f(x_1)}{x_2-x_1}=\dfrac{f(8)-f(-4)}{8-(-4)}=\dfrac{((8)^2+8(8)-5)-((-4)^2+8(-4)-5)}{8-(-4)}$

$$=\dfrac{123+21}{12}=\dfrac{144}{12}=12$$

214. $\dfrac{f(x_2)-f(x_1)}{x_2-x_1}=\dfrac{f(d)-f(c)}{d-c}=\dfrac{(m(d)+b)-(m(c)+b)}{d-c}$

$$=\dfrac{md-mc}{d-c}=\dfrac{m(d-c)}{d-c}=m$$

215. $\dfrac{f(x_2)-f(x_1)}{x_2-x_1}=\dfrac{f(n)-f(m)}{n-m}=\dfrac{(an^2+bn+c)-(am^2+bm+c)}{n-m}$

$$=\dfrac{an^2+bn-am^2-bm}{n-m}=\dfrac{an^2-am^2+bn-bm}{n-m}$$

$$=\dfrac{a(n^2-m^2)+b(n-m)}{n-m}=\dfrac{a(n-m)(n+m)+b(n-m)}{(n-m)}$$

$$=a(n+m)+b$$

216. $\dfrac{f(x+h)-f(x)}{h}=\dfrac{(2(x+h)-5)-(2x-5)}{h}=\dfrac{2x+2h-5-2x+5}{h}=\dfrac{2h}{h}=2$

217. $\dfrac{g(x+h)-g(x)}{h}=\dfrac{(2(x+h)+50)-(2x+50)}{h}=\dfrac{2x+2h+50-2x-50}{h}=\dfrac{2h}{h}=2$

218. $\dfrac{f(x+h)-f(x)}{h}=\dfrac{((x+h)^2+3(x+h))-(x^2+3x)}{h}$

$$=\dfrac{(x^2+2xh+h^2+3x+3h)-(x^2+3x)}{h}=\dfrac{2xh+h^2+3h}{h}$$

$$=2x+h+3$$

219. $\dfrac{f(x+h)-f(x)}{h}=\dfrac{(a(x+h)^2+b(x+h)+c)-(ax^2+bx+c)}{h}$

$$=\dfrac{(a(x^2+2xh+h^2)+b(x+h)+c)-(ax^2+bx+c)}{h}$$

$$=\dfrac{(ax^2+2axh+ah^2+bx+bh+c)-ax^2-bx-c}{h}$$

$$=\dfrac{2axh+ah^2+bh}{h}=2ax+ah+b$$

220. $\dfrac{f(x+h)-f(x)}{h} = \dfrac{(m(x+h)+b)-(mx+b)}{h} = \dfrac{mx+mh+b-mx-b}{h} = \dfrac{mh}{h} = m$

221. (A) $f(t)=65t$; the output is in miles $\left(\dfrac{\text{mi}}{\text{hr}} \cdot \text{hr} = \text{mi}\right)$

   (B) $f\left(2\dfrac{3}{5}\right) = 65\left(2\dfrac{3}{5}\right) = 169$ miles

222. (A) $f(t)=2{,}000-150t$; the output is in gallons

   (B) $f(3)=2{,}000-150(3)=2{,}000-450=1{,}550$ gallons

223. (A) $f(t)=245-60t$; the output is in miles

   (B) $f(2.5)=245-60(2.5)=95$ miles

224. (A) $F(x)=-24{,}000x$; the output is in dynes

   (B) $F(4)=-24{,}000(4)=-96{,}000$ dynes

225. (A) $f(x)=121+0.555x$; the output is in dollars

   (B) $f(180)=121+0.555(180)=\$220.90$

226. (A) The perimeter of the fence is $480 = 2W + 2L$; thus, $L = 240 - W$.

   The area of the rectangular pen equals length times width, so

   $$f(W) = (240 - W)W = 240W - W^2 = -W^2 + 240W.$$

   (B) The graph of $f(W)$ is a parabola opening downward. The vertex formula is $\left(-\dfrac{b}{2a}, f\left(-\dfrac{b}{2a}\right)\right)$, so the maximum value for $f(W)$ occurs when $W = -\dfrac{b}{2a} = -\dfrac{240}{2(-1)} = 120$ feet. The corresponding length is $L = 240 - W = 240 - 120 = 120$ feet. Therefore, the dimensions that maximize the area are 120 feet by 120 feet.

227. The graph of $h(t)$ is a parabola opening downward. The vertex formula is $\left(-\dfrac{b}{2a}, h\left(-\dfrac{b}{2a}\right)\right)$, so the maximum value for $h(t)$ occurs when $t = -\dfrac{b}{2a} = -\dfrac{100}{2(-16)} = 3.125$ seconds. Therefore, the maximum height is reached at $t = 3.125$ seconds: $h(3.125) = -16(3.125)^2 + 100(3.125) + 30 = 186.25$ feet.

228. The graph of $R(x)$ is a parabola opening downward. The vertex formula is $\left(-\dfrac{b}{2a}, R\left(-\dfrac{b}{2a}\right)\right)$, so the maximum value for $R(x)$ occurs when $x = -\dfrac{b}{2a} = -\dfrac{160}{2(-4)} =$ 20 thousand boxes of designer watches. Therefore, the maximum revenue is reached at $x = 20$: $R(20) = -4(20)^2 + 160(20) = 1600$ thousands of dollars $= \$1{,}600{,}000$.

229. The graph of $C(x)$ is a parabola opening upward. The vertex formula is $\left(-\dfrac{b}{2a}, C\left(-\dfrac{b}{2a}\right)\right)$, so the minimum value for $C(x)$ occurs when $x = -\dfrac{b}{2a} = -\dfrac{-56}{2(1)} = 28$. Therefore, the minimum yearly cost is reached at $x = 28$: $C(28) = (28)^2 - 56(28) + 3{,}000 = 2{,}216$ thousands of dollars $= \$2{,}216{,}000$.

230. The graph of $V(t)$ is a parabola opening downward. The vertex formula is $\left(-\dfrac{b}{2a}, V\left(-\dfrac{b}{2a}\right)\right)$, so the maximum value for $V(t)$ occurs when $t = -\dfrac{b}{2a} = -\dfrac{10}{2(-1)} = 5$ hundred years.

Chapter 5: Polynomial and Rational Functions

231. (a) degree $= 5$; (b) domain $= R$, range $= R$; (c) zeros: $x = -3, -2, 1, 2, 4$; (d) x-intercepts $= -3, -2, 1, 2, 4$; (e) y-intercept $= p(0) = 3(0-1)(0+3)(0-4)(0+2)(0-2) = -144$

232. (a) degree $= 6$; (b) domain $= R$, range $\subset R$; (c) $q(x) = (x^2 + 4)(x^2 - 5)(x^2 - 9) = (x + 2i)(x - 2i)(x + \sqrt{5})(x - \sqrt{5})(x + 3)(x - 3)$; zeros: $x = -3, -\sqrt{5}, \sqrt{5}, 3, -2i, 2i$; (d) x-intercepts $= -3, -\sqrt{5}, \sqrt{5}, 3$; (e) y-intercept $= q(0) = (0^2 + 4)(0^2 - 5)(0^2 - 9) = 180$

233. (a) degree $= 4$; (b) domain $= R$, range $= [-81, \infty)$; (c) $g(x) = x^4 - 81 = (x^2 + 9)(x^2 - 9) = (x + 3i)(x - 3i)(x + 3)(x - 3)$; zeros: $x = -3, 3, -3i, 3i$; (d) x-intercepts $= -3, 3$; (e) y-intercept $= g(0) = 0^4 - 81 = -81$

234. (a) degree $= 2$; (b) domain $= R$, range $= (-\infty, 15.125]$ *Note:* The graph of g is a parabola that turns downward with vertex $(-2.25, 15.125)$; (c) $g(x) = -2x^2 - 9x + 5 = -(2x^2 + 9x - 5) = -(2x - 1)(x + 5)$; zeros: $x = -5, \dfrac{1}{2}$; (d) x-intercepts $= -5, \dfrac{1}{2}$; (e) y-intercept $= g(0) = -2 \cdot 0^2 - 9 \cdot 0 + 5 = 5$

235. (a) degree $= 1$; (b) domain $= R$, range $= R$; (c) zero: $x = -\dfrac{5}{3}$; (d) x-intercept $= -\dfrac{5}{3}$; (e) y-intercept $= f(0) = 3 \cdot 0 + 5 = 5$

236. (a) turning points: $(-7.22, 40.31)$, $(-2.42, -22.76)$, $(1.97, 3.41)$, $(5.27, -17.68)$; (b) relative maxima: 40.31, 3.41; relative minima: -22.76, -17.68; no absolute extrema

237. The remainder theorem states that if a polynomial $p(x)$ is divided by $x - a$, the remainder is $p(a)$. Using synthetic division,

$$\begin{array}{r} 2\,\big|\,2 \quad -5 \quad -14 \quad 8 \\ \underline{\quad\quad 4 \quad -2 \quad -32} \\ 2 \quad -1 \quad -16 \quad \big|\underline{-24}\ \text{Remainder} \end{array}$$

Thus, $p(2) = -24$.

238. Using synthetic division,

$$\begin{array}{r} -2\,\big|\,2 \quad -5 \quad -14 \quad 8 \\ \underline{\quad\quad -4 \quad 18 \quad -8} \\ 2 \quad -9 \quad 4 \quad \big|\underline{0}\ \text{Remainder} \end{array}$$

Thus, $p(-2) = 0$.

239. The factor theorem states that $x - c$ is a factor of $p(x)$ if and only if $p(c) = 0$. Therefore, because $p(-2) = 0$, $(x + 2)$ is a factor of $p(x)$. From the synthetic division in question 238 $p(x) = (x + 2)(2x^2 - 9x + 4)$. Factoring the quadratic yields $p(x) = (x + 2)(2x^2 - 9x + 4) = (x + 2)(2x - 1)(x - 4)$.

240. $p(x) = 5(x - 3)(x + 2)(x + \sqrt{2})(x - \sqrt{2})$

241. $g(x) = 2(x + 1)(x - 1)(x - 3)$

242. (A) one
(B) k
(C) roots
(D) n
(E) $x - yi$

243. (a) $p(x) = (x + 3)(x^2 - 5)(x^2 - 49)(x^2 + 49)$

$$= (x + 3)(x + \sqrt{5})(x - \sqrt{5})(x + 7)(x - 7)(x + 7i)(x - 7i)$$

(b) zeros: $7i, -7i, -7, -3, -\sqrt{5}, \sqrt{5}, 7$

244. (a) $p(x) = (3x^2 - x - 10)(x^6 - 64) = (3x + 5)(x - 2)(x^3 + 8)(x^3 - 8)$

$= (3x + 5)(x - 2)(x + 2)(x^2 - 2x + 4)(x - 2)(x^2 + 2x + 4)$

$= (3x + 5)(x - 2)(x + 2)(x + 1 + i\sqrt{3})(x + 1 - i\sqrt{3})(x - 2)(x - 1 + i\sqrt{3})(x - 1 - i\sqrt{3})$

$= (3x + 5)(x - 2)^2(x + 2)(x + 1 + i\sqrt{3})(x + 1 - i\sqrt{3})(x - 1 + i\sqrt{3})(x - 1 - i\sqrt{3})$

(b) zeros: $-\dfrac{5}{3}$, 2 (multiplicity 2), $-2, 1 + i\sqrt{3}, 1 - i\sqrt{3}, -1 + i\sqrt{3}, -1 - i\sqrt{3}$

245. The polynomial has real coefficients; so another root is $2 + i$, the complex conjugate of $2 - i$. Each root corresponds to a factor of the polynomial yielding

$p(x) = (x - 3)[x - (2 - i)][x - (2 + i)]$

$p(x) = (x - 3)[(x - 2) + i][(x - 2) - i]$ Regroup.

$p(x) = (x - 3)[(x - 2)^2 - i^2]$ Multiply last two factors.

$p(x) = (x - 3)[(x^2 - 4x + 4) - (-1)]$ Simplify.

$p(x) = x^3 - 7x^2 + 17x - 15$

This polynomial equation has the required roots and the least degree.

246. The number of sign variations in $p(x) = 6x^4 + 7x^3 - 9x^2 - 7x + 3$ is 2, so $p(x) = 0$ has 2 or 0 positive real roots. The number of sign variations in $p(-x) = 6x^4 - 7x^3 - 9x^2 + 7x + 3$ is 2, so $p(x) = 0$ has 2 or 0 negative real roots.

247. The number of sign variations in $p(x) = x^3 - 1$ is 1, so $p(x) = 0$ has at most one positive real root. The number of sign variations in $p(-x) = -x^3 - 1$ is 0, so $p(x) = 0$ has no negative real roots.

248. The number of sign variations in $p(x) = x^5 + 4x^4 - 4x^3 - 16x^2 + 3x + 12$ is 2, so $p(x) = 0$ has 2 or 0 positive real roots. The number of sign variations in $p(-x) = -x^5 + 4x^4 + 4x^3 - 16x^2 - 3x + 12$ is 3, so $p(x) = 0$ has 3 or 1 negative real roots.

249. The rational root theorem states that if $a_n x^n + a_{n-1}x^{n-1} + a_{n-2}x^{n-2} + \ldots + a_2 x^2 + a_1 x + a_0 = 0$ is a polynomial equation with *integral* coefficients and it has a rational

root $\dfrac{p}{q}$ (in lowest terms), then p is a factor of a_0 and q is a factor of a_n.

Possible numerators of a rational zero are factors of 6: $\pm 1, \pm 2, \pm 3$, and ± 6. Possible denominators are factors of 2: ± 1 and ± 2. Thus, possible rational roots of

$2x^3 + x^2 - 13x + 6 = 0$ are $\pm 1, \pm 2, \pm 3, \pm 6, \pm\dfrac{1}{2}$, and $\pm\dfrac{3}{2}$.

250. The equation has degree 3, so it has at most three roots. The number of sign variations in $p(x) = 2x^3 + x^2 - 13x + 6$ is 2, so $p(x) = 0$ has 2 or 0 positive real roots. The number of sign variations in $p(-x) = -2x^3 + x^2 + 13x + 6$ is 1, so $p(x) = 0$ at most 1 negative real root. From question 249, the possible rational roots of $p(x) = 0$ are ±1, ±2, ±3, ±6, $\pm\dfrac{1}{2}$, and $\pm\dfrac{3}{2}$. Use the factor theorem to test ±1:

$p(1) = 2 \cdot 1^3 + 1^2 - 13 \cdot 1 + 6 = -4 \neq 0$, so 1 is not a zero of p and, therefore, not a root of $p(x) = 0$.
$p(-1) = 2 \cdot (-1)^3 + (-1)^2 - 13 \cdot (-1) + 6 = 18 \neq 0$, so -1 is not a zero of p and, therefore, not a root of $p(x) = 0$.

Use the remainder theorem to test the other possible roots.

Test $x = 2$:

$$
\begin{array}{r|rrr}
2 & 2 & 1 & -13 & 6 \\
  &   & 4 & 10 & -6 \\
\hline
  & 2 & 5 & -3 & \underline{|0}\ \text{Remainder}
\end{array}
$$

Thus, 2 is a zero of p and, therefore, a root of $p(x) = 0$.

Hence, by the factor theorem, $(x - 2)$ is a factor of $p(x)$. Using the coefficients from the synthetic division, $p(x) = 2x^3 + x^2 - 13x + 6 = (x - 2)(2x^2 + 5x - 3)$. Factoring completely gives $p(x) = (x - 2)(2x - 1)(x + 3)$. Thus, the roots of $p(x) = 0$ are 2, $\dfrac{1}{2}$, and -3.

251. (a) domain $= \{x \mid x \neq 0\}$; (b) zeros: none; (c) x-intercepts: none; (d) y-intercepts: none

252. $g(x) = \dfrac{x^2 - 36}{3x^2 - x - 10} = \dfrac{(x + 6)(x - 6)}{(3x + 5)(x - 2)}$ (a) domain $= \left\{x \,\middle|\, x \neq -\dfrac{5}{3}, 2\right\}$;

(b) zeros: $x = -6, 6$; (c) x-intercepts $= -6, 6$; (d) y-intercept $= g(0) = \dfrac{0^2 - 36}{3 \cdot 0^2 - 0 - 10} = \dfrac{18}{5}$

253. $f(x) = \dfrac{x^2 - 25}{x^2 - x - 6} = \dfrac{(x + 5)(x - 5)}{(x + 2)(x - 3)}$ (a) domain $= \{x \mid x \neq -2, 3\}$;

(b) zeros: $x = -5, 5$; (c) x-intercepts $= -5, 5$; (d) y-intercept $= f(0) = \dfrac{0^2 - 25}{0^2 - 0 - 6} = \dfrac{25}{6}$

254. $f(x) = \dfrac{x^3 - 8}{x - 4} = \dfrac{(x - 2)(x^2 + 2x + 4)}{x - 4}$ (a) domain $= \{x \mid x \neq 4\}$;

(b) zeros: $x = 2, -1 + i\sqrt{3}, -1 - i\sqrt{3}$; (c) x-intercept $= 2$; (d) y-intercept $= f(0) = \dfrac{0^3 - 8}{0 - 4} = 2$

255. $f(x) = \dfrac{x^3 - 8}{x^2 + 2x + 4} = \dfrac{(x - 2)(x^2 + 2x + 4)}{(x^2 + 2x + 4)} = x - 2$ (a) domain $= R$; (b) zeros: $x = 2$;

(c) x-intercept $= 2$; (d) y-intercept $= f(0) = 0 - 2 = -2$

256. $h(x) = \dfrac{x+1}{x^2-1} = \dfrac{(x+1)}{(x+1)(x-1)} = \dfrac{1}{(x-1)}$ (a) domain = $\{x \mid x \neq -1,1\}$; (b) zero: none

(because $x = -1$ is not in the domain); (c) x-intercepts = none; (d) y-intercept = $h(0) =$

$\dfrac{1}{0-1} = -1$

257. (a) vertical asymptote: $x = 0$; (b) holes: none

258. $g(x) = \dfrac{x^2-4}{3x^2-x-10} = \dfrac{(x+2)(x-2)}{(3x+5)(x-2)} = \dfrac{x+2}{3x+5}$ (a) vertical asymptote: $x = -\dfrac{5}{3}$;

(b) hole: $\left(2, \dfrac{4}{11}\right)$

259. $f(x) = \dfrac{x^3-8}{x-4} = \dfrac{(x-2)(x^2+2x+4)}{x-4}$ (a) vertical asymptote: $x = 4$; (b) holes: none

260. $f(x) = \dfrac{x^3-8}{x^2+2x+4} = \dfrac{(x-2)(x^2+2x+4)}{(x^2+2x+4)} = x-2$ (a) vertical asymptotes: none;

(b) holes: none

261. $h(x) = \dfrac{x+3}{x^2-9} = \dfrac{(x+3)}{(x+3)(x-3)} = \dfrac{1}{(x-3)}$ (a) vertical asymptote: $x = 3$;

(b) hole: $\left(-3, -\dfrac{1}{6}\right)$

262. The numerator's degree is less than the denominator's degree, so the x-axis is a horizontal asymptote.

263. $g(x) = \dfrac{x^2-4}{3x^2-x-10} = \dfrac{(x+2)(x-2)}{(3x+5)(x-2)} = \dfrac{x+2}{3x+5}$. The numerator's degree equals the

denominator's degree, so $y = \dfrac{1}{3}$ is a horizontal asymptote.

264. $f(x) = \dfrac{x^3-8}{x-4}$ is in simplified form. The numerator's degree exceeds the denomina-

tor's degree by more than 1, so there is no horizontal or oblique asymptote.

265. $f(x) = \dfrac{x^2 + 3}{x - 1}$ is in simplified form. The numerator's degree exceeds the denominator's

degree by exactly 1, so the graph has an oblique asymptote. $f(x) = \dfrac{x^2 + 3}{x - 1} = x + 1 + \dfrac{4}{x - 1}$;

hence, $y = x + 1$ is an oblique asymptote of the graph of f.

266. $h(x) = \dfrac{x + 3}{x^2 - 9} = \dfrac{(x + 3)}{(x + 3)(x - 3)} = \dfrac{1}{(x - 3)}$. The numerator's degree is less than the

denominator's degree, so the x-axis is a horizontal asymptote.

267. As x approaches ∞ or $-\infty$, $f(x)$ behaves as does x^5. Therefore, (a) as x approaches ∞, $f(x)$ approaches ∞ because x^5 approaches ∞; (b) as x approaches $-\infty$, $f(x)$ approaches $-\infty$ because x^5 approaches $-\infty$.

268. As x approaches ∞ or $-\infty$, $f(x)$ behaves as does $6x^4$. Therefore, (a) as x approaches ∞, $f(x)$ approaches ∞ because $6x^4$ approaches ∞; (b) as x approaches $-\infty$, $f(x)$ approaches ∞ because $6x^4$ approaches ∞.

269. As x approaches ∞ or $-\infty$, $f(x)$ behaves as does $-2x^4$. Therefore, (a) as x approaches ∞, $f(x)$ approaches $-\infty$ because $-2x^4$ approaches $-\infty$; (b) as x approaches $-\infty$, $f(x)$ approaches $-\infty$ because $-2x^4$ approaches $-\infty$.

270. As x approaches ∞ or $-\infty$, $f(x)$ behaves as does $\dfrac{3x}{x^2}$. Therefore, (a) as x approaches ∞,

$f(x)$ approaches 0 because $\dfrac{3x}{x^2}$ approaches 0; (b) as x approaches $-\infty$, $f(x)$ approaches 0

because $\dfrac{3x}{x^2}$ approaches 0.

271. As x approaches ∞ or $-\infty$, $f(x)$ behaves as does $\dfrac{10x^2}{5x^2}$. Therefore, (a) as x approaches

∞, $f(x)$ approaches 2 because $\dfrac{10x^2}{5x^2}$ approaches 2; (b) as x approaches $-\infty$, $f(x)$ approaches

2 because $\dfrac{10x^2}{5x^2}$ approaches 2.

272. As x approaches ∞ or $-\infty$, $f(x)$ behaves as does $\dfrac{2x^3}{-5x^2}$. Therefore, (a) as x approaches

∞, $f(x)$ approaches $-\infty$ because $\dfrac{2x^3}{-5x^2}$ approaches $-\infty$; (b) as x approaches $-\infty$, $f(x)$

approaches ∞ because $\dfrac{2x^3}{-5x^2}$ approaches ∞.

Chapter 6: Exponential, Logarithmic, and Other Common Functions

273. $f(4) = 5^4 = 625$

274. $f\left(\dfrac{4}{3}\right) = 27^{\frac{4}{3}} = (\sqrt[3]{27})^4 = 3^4 = 81$

275. $f(-6) = \left(\dfrac{1}{2}\right)^{-6} = 2^6 = 64$

276. (a) The domain is R and the range is $(0,\infty)$; (b) There are no zeros; (c) $x = 0$ is a horizontal asymptote; (d) The y-intercept is $y = 1$ and there are no x-intercepts; (e) $b = 3 > 1$, therefore f is increasing on R; (f) As x approaches ∞, $f(x)$ approaches ∞ and as x approaches $-\infty$, $f(x)$ approaches 0.

277. (a) The domain is R and the range is $(0,\infty)$; (b) There are no zeros; (c) $x = 0$ is a horizontal asymptote; (d) The y-intercept is 1 and there are no x-intercepts; (e) $b = \dfrac{1}{4} < 1$, therefore f is decreasing on R; (f) As x approaches ∞, $f(x)$ approaches 0 and as x approaches $-\infty$, $f(x)$ approaches ∞.

278. $\dfrac{f(8)}{f(3)} = \dfrac{e^8}{e^3} = e^{8-3} = e^5$

279. $f(3) \cdot f(2) = 2^3 \cdot 2^2 = 2^{3+2} = 2^5 = 32$

280. $f(-1) = \left(\dfrac{7}{8}\right)^{-1} = \dfrac{8}{7}$

281. Using the one-to-one property, $u = 21$.

282. Logarithms are ways to write exponents. If $\log_b x = k$, then k is the *exponent* that is used on b to get x; that is, $x = b^k$. Therefore, $f(625) = \log_5 625 = 4$ because $5^4 = 625$.

283. $f(81) = \log_{27} 81 = \dfrac{4}{3}$ because $27^{\frac{4}{3}} = (\sqrt[3]{27})^4 = 3^4 = 81$.

284. $f(64) = \log_{\frac{1}{2}} 64 = -6$ because $\left(\dfrac{1}{2}\right)^{-6} = 2^6 = 64$.

285. $h\left(\dfrac{8}{27}\right) = \log_{\frac{4}{9}}\left(\dfrac{8}{27}\right) = \dfrac{3}{2}$ because $\left(\dfrac{4}{9}\right)^{\frac{3}{2}} = \left(\sqrt{\dfrac{4}{9}}\right)^3 = \left(\dfrac{2}{3}\right)^3 = \dfrac{8}{27}$.

286. (A) The exponential function $g(x) = b^x$ is the inverse of the logarithmic function $f(x) = \log_b x$ ($b \neq 1, b > 0$). Thus, $g(x) = 6^x$ is the inverse of $f(x) = \log_6 x$.

(B) The logarithmic function $f(x) = \log_b x$ is the inverse of the exponential function $g(x) = b^x$ ($b \neq 1, b > 0$). Thus, $g(x) = \log_{1.035} x$ is the inverse of $f(x) = (1.035)^x$.

(C) $g(x) = \log_{\frac{3}{4}} x$ is the inverse of $f(x) = \left(\dfrac{3}{4}\right)^x$.

(D) $f(x) = \ln x$ is the natural logarithmic function that has base e; that is $f(x) = \ln x = \log_e x$, where e is the irrational constant approximately equal to 2.718281828. Thus, its inverse is $g(x) = e^x$. See the following figure.

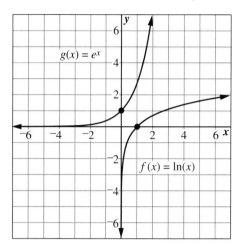

(E) $f(x) = \log x$ is the common logarithmic function that has base 10; that is $f(x) = \log x = \log_{10} x$. Thus, its inverse is $g(x) = 10^x$.

287. (a) The domain is $(0,\infty)$ and the range is $(-\infty,\infty)$; (b) $x = 1$ is the only zero; (c) The y-axis is a vertical asymptote; (d) The x-intercept is $(1,0)$ and there are no y-intercepts; (e) $b = 6 > 1$, so $f(x)$ is increasing on $(0,\infty)$; (f) $b = 6 > 1$, so as x approaches 0, $f(x)$ approaches $-\infty$ and as x approaches ∞, $f(x)$ approaches ∞.

288. (a) The domain is $(0,\infty)$ and the range is $(-\infty,\infty)$; (b) $x = 1$ is the only zero; (c) The y-axis is a vertical asymptote; (d) The x-intercept is $(1,0)$ and there are no y-intercepts; (e) $b = \dfrac{1}{2} < 1$, so $f(x)$ is decreasing on $(0,\infty)$; (f) $b = \dfrac{1}{2} < 1$, so as x approaches 0, $f(x)$ approaches ∞ and as x approaches ∞, $f(x)$ approaches $-\infty$.

289. $f(e^{10}) = \ln e^{10} = 10 \ln e = 10 \cdot 1 = 10$

290. $g(64^{20}) = \log_2 64^{20} = 20 \log_2 64 = 20(6) = 120$

291. $h\left(\dfrac{100}{0.000001}\right) = \log\left(\dfrac{100}{0.000001}\right) = \log(100) - \log(0.000001) = \log(10^2) - \log(10^{-6}) = 2 - (-6) = 2 + 6 = 8$

292. $g(8 \cdot 32) = \log_2(8 \cdot 32) = \log_2 8 + \log_2 32 = 3 + 5 = 8$

293. Using the one-to-one property, $u = 450$.

294. (A) $\log_8(32,768) = \dfrac{\ln 32,768}{\ln 8} = 5$

(B) $\log_{\frac{1}{5}}(0.0016) = \dfrac{\ln(0.0016)}{\ln\left(\dfrac{1}{5}\right)} = 4$

(C) $\log_2(4,096) = \dfrac{\ln(4,096)}{\ln 2} = 12$

(D) $\log_{1.05}(2.5) = \dfrac{\ln(2.5)}{\ln(1.05)} \approx 18.78$

(E) $\log_2(400) = \dfrac{\ln 400}{\ln 2} \approx 8.64$

295. $8\log_2(3x-1) = 256$

$\log_2(3x-1) = 32$	Isolate the $\log_2$ function: divide both sides by 8.
$2^{\log_2(3x-1)} = 2^{32}$	Undo $\log_2$: exponentiate both sides, base 2.
$3x - 1 = 2^{32}$	Use inverse property $b^{\log_b x} = x$: $2^{\log_2(3x-1)} = 3x - 1$.
$3x = 2^{32} + 1$	Isolate the term containing x: add 1 to both sides.
$x = \dfrac{2^{32}+1}{3}$	Solve for x: divide both sides by 3.
$x = 1,431,655,766$	Evaluate.

296. $\ln 8x = 3.5$

$e^{\ln 8x} = e^{3.5}$	Undo ln: exponentiate both sides, base e.
$8x = e^{3.5}$	Use inverse property $e^{\ln x} = x$: $e^{\ln 8x} = 8x$.
$x = \dfrac{e^{3.5}}{8}$	Solve for x: divide both sides by 8.
$x = \dfrac{e^{3.5}}{8}$	Solve for x: divide both sides by 8.
$x \approx 4.14$	Approximate.

297. $4000(1.005)^x = 12,000$

$(1.005)^x = 3$	Isolate the exponential function: divide both sides by 4000.
$\ln(1.005)^x = \ln 3$	Undo exponentiation: take the natural log of both sides.
$x\ln(1.005) = \ln 3$	Use property $\log_b(u^p) = p\log_b u$: $\ln(1.005)^x = x\ln(1.005)$.
$x = \dfrac{\ln 3}{\ln(1.005)}$	Solve for x: divide both sides by $\ln(1.005)$.
$x \approx 220.27$	Approximate.

298. $75e^{0.05x} = 150$

$e^{0.05x} = 2$ Isolate the exponential function: divide both sides by 75.

$\ln e^{0.05x} = \ln 2$ Undo exponentiation: take the natural log of both sides.

$0.05x = \ln 2$ Use inverse property $\ln e^x = x$: $\ln e^{0.05x} = 0.05x$.

$x = \dfrac{\ln 2}{0.05}$ Solve for x: divide both sides by 0.05.

$x \approx 13.86$ Approximate.

299. $e^{6x+1} = e^{4x-4}$

$6x + 1 = 4x - 4$ Use the one-to-one property.

$2x = -5$ Isolate the variable term.

$x = -\dfrac{5}{2}$ Solve for x.

300. Substituting $A_0 = 20$ and $k = 5{,}730$ into the formula, $A(t) = A_o \left(\dfrac{1}{2}\right)^{\frac{t}{k}}$, and evaluating

at $t = 5{,}000$, omitting units for convenience, yields $A(5{,}000) = 20 \left(\dfrac{1}{2}\right)^{\frac{5000}{5730}} \approx 10.92$ grams.

301. Evaluating the formula for $x = 7.6 \times 10^{-4}$ yields pH of diet soda

$Z = f(7.6 \times 10^{-4}) = -\log_{10}(7.6 \times 10^{-4}) \approx 3.12$.

302. To find the rate, compounded annually, that will double an investment of $50,000 in 20 years, substitute $P = \$100{,}000$, $P_0 = \$50{,}000$, and $t = 20$ years into the formula (omitting the units) and then solve for r.

$$P = P_0(1+r)^t$$
$$100{,}000 = 50{,}000(1+r)^{20}$$
$$2 = (1+r)^{20}$$
$$\ln 2 = \ln(1+r)^{20}$$
$$\ln 2 = 20\ln(1+r)$$
$$\frac{\ln 2}{20} = \ln(1+r)$$
$$e^{\frac{\ln 2}{20}} = 1+r$$
$$1+r = e^{\frac{\ln 2}{20}}$$
$$r = e^{\frac{\ln 2}{20}} - 1$$
$$r \approx .035 = 3.5\%$$

303. $f(x) = \begin{cases} 2x+5 & \text{if } x < 0 \\ 3 & \text{if } x = 0 \\ \sqrt{2x+5} & \text{if } x > 0 \end{cases}$ is a *piecewise function*. A piecewise function is a function

defined by different equations on different parts (usually intervals) of its domain. Evaluate the function at x according to the interval in which x falls.

(A) Because $x = -1 < 0$, $f(-1) = 2(-1) + 5 = 3$.

(B) Because $x = -\dfrac{5}{2} < 0$, $f\left(-\dfrac{5}{2}\right) = 2\left(-\dfrac{5}{2}\right) + 5 = 0$.

(C) Because $x = 0$, $f(0) = 3$.

(D) Because $x = 5.5 > 0$, $f(5.5) = \sqrt{2(5.5)+5} = \sqrt{16} = 4$.

(E) Because $x = 3 > 0$, $f(3) = \sqrt{2(3)+5} = \sqrt{11}$.

304. $f(x) = [x] = $ the greatest integer n such that $n \leq x$.

(A) $f(3.99) = [3.99] = 3$

(B) $f(-3.99) = [-3.99] = -4$

(C) $f(0.005) = [0.005] = 0$

(D) $f(-15) = [-15] = -15$

(E) $f(\pi) = [\pi] = 3$

305. $\sqrt{x^2} = 9$

$|x| = 9$

$x = 9 \text{ or } x = -9$

Solution set: $\{-9, 9\}$

306. $|-2x - 3| < 15$

$|-(2x+3)| < 15$

$|2x+3| < 15$

$-15 < 2x + 3 < 15$

$-15 - 3 < 2x + 3 - 3 < 15 - 3$

$-18 < 2x < 12$

$\dfrac{-18}{2} < \dfrac{2x}{2} < \dfrac{12}{2}$

$-9 < x < 6$

Solution set: $(-9, 6)$

307. $|5x| \geq 40$

$|5||x| \geq 40$

$5|x| \geq 40$

$\dfrac{5|x|}{5} \geq \dfrac{40}{5}$

$|x| \geq 8$

$x \leq -8 \text{ or } x \geq 8$

Solution set: $(-\infty, -8] \cup [8, \infty)$

308. $-2|4x + 1| \geq -10$

$|4x + 1| \leq 5$ *Note*: Don't forget to reverse the inequality.

$-5 \leq 4x + 1 \leq 5$

$-5 - 1 \leq 4x + 1 - 1 \leq 5 - 1$

$-6 \leq 4x \leq 4$

$-\dfrac{3}{2} \leq x \leq 1$

Solution set: $\left[-\dfrac{3}{2}, 1\right]$

309. Solution set: $\varnothing$, because $|x| \geq 0$

310. $f(10) = 10^{-3} = \dfrac{1}{10^3} = \dfrac{1}{1{,}000}$

311. $f(256) = 256^{0.75} = 64$

312. $f(6) = 6^{\sqrt{2}} \approx 12.60$

Chapter 7: Matrices and Systems of Linear Equations

313. The *rows* of A are its horizontal lines of entries, so A has three rows. The *columns* of A are its vertical lines of entries, so A has two columns. The *size* (also called *order*) is 3×2, its number of rows by its number of columns.

314. Recall that the element a_{ij} is in the ith row and jth column of matrix A.

(A) $a_{13} = 4$

(B) $a_{31} = -4$

(C) $a_{35} = -7$

(D) $a_{14} = 3$

(E) $a_{33} = 8$

315. (A) A column vector is a matrix with only one column, Thus, the 3×1 column vector

whose elements are 2, 3, and 8 is $\begin{bmatrix} 2 \\ 3 \\ 8 \end{bmatrix}$.

(B) A row vector is a matrix with only one row. Thus, the 1×4 row vector whose elements are 0, -2, 1, 5 is $\begin{bmatrix} 0 & -2 & 1 & 5 \end{bmatrix}$.

(C) For any square matrix $A = [a_{ij}]_{n \times n}$, the main diagonal elements are $a_{11}, a_{22}, a_{33}, \ldots, a_{nn}$. *Note*: Hereafter a matrix's main diagonal will be called simply its diagonal. Thus, the 3×3 matrix $A = [a_{ij}]_{3 \times 3}$ whose diagonal elements are 3, -4, 5

and whose off-diagonal elements are 1's is $\begin{bmatrix} 3 & 1 & 1 \\ 1 & -4 & 1 \\ 1 & 1 & 5 \end{bmatrix}$.

(D) $I_{3 \times 3}$ denotes the 3×3 identity matrix. An identity matrix is a square matrix whose diagonal elements are 1's and whose off-diagonal elements are 0's. Thus, $I_{3 \times 3} = \begin{bmatrix} 1 & 0 & 0 \\ 0 & 1 & 0 \\ 0 & 0 & 1 \end{bmatrix}$.

(E) All the elements of the $m \times n$ zero matrix, denoted 0 (or $0_{m \times n}$), are zero. Thus, the 2×3 zero matrix is $\begin{bmatrix} 0 & 0 & 0 \\ 0 & 0 & 0 \end{bmatrix}$.

316. Matrices can be added (subtracted) only if they are the same size. The sum (difference) of two matrices is the matrix whose elements are the sums (differences) of the corresponding elements of the two matrices.

$$\begin{bmatrix} 4 & 10 \\ -3 & 0 \end{bmatrix} + \begin{bmatrix} 3 & 2 \\ -2 & 3 \end{bmatrix} = \begin{bmatrix} 4+3 & 10+2 \\ -3-2 & 0+3 \end{bmatrix} = \begin{bmatrix} 7 & 12 \\ -5 & 3 \end{bmatrix}$$

317. Refer to question 316.

$$\begin{bmatrix} 3 & 6 & 0 \\ 8 & -2 & -5 \\ 6 & -4 & 0 \end{bmatrix} - \begin{bmatrix} -1 & 6 & 4 \\ 0 & 3 & -1 \\ 5 & 0 & 3 \end{bmatrix} = \begin{bmatrix} 3-(-1) & 6-(6) & 0-(4) \\ 8-(0) & -2-(3) & -5-(-1) \\ 6-(5) & -4-(0) & 0-(3) \end{bmatrix} = \begin{bmatrix} 4 & 0 & -4 \\ 8 & -5 & -4 \\ 1 & -4 & -3 \end{bmatrix}$$

318. For a matrix $A = [a_{ij}]_{m \times n}$ and a scalar k, $kA = [ka_{ij}]_{m \times n}$ is the product A by k. Thus,

$$2 \begin{bmatrix} 3 & 1 & 1 \\ 1 & -4 & 1 \\ 1 & 1 & 5 \end{bmatrix} = \begin{bmatrix} 2 \cdot 3 & 2 \cdot 1 & 2 \cdot 1 \\ 2 \cdot 1 & 2 \cdot -4 & 2 \cdot 1 \\ 2 \cdot 1 & 2 \cdot 1 & 2 \cdot 5 \end{bmatrix} = \begin{bmatrix} 6 & 2 & 2 \\ 2 & -8 & 2 \\ 2 & 2 & 10 \end{bmatrix}.$$

319. The product of a $1 \times n$ row vector and a $n \times 1$ column vector is obtained by multiplying corresponding elements and adding the resulting products. Thus,

$$[3 \quad 6 \quad 1] \begin{bmatrix} 2 \\ -3 \\ 5 \end{bmatrix} = [(3)(2) + (6)(-3) + (1)(5)] = [6 - 18 + 5] = [-7].$$

Note: The product is undefined if the row and column vectors do not have the same number of elements.

320. If $A = [a_{ij}]_{m \times k}$ is a matrix of size $m \times k$ and $B = [b_{ij}]_{k \times n}$ is a matrix of size $k \times n$ so that the number of columns of A is equal to the number of rows of B, then the product AB is the $m \times n$ matrix $C = [c_{ij}]_{m \times n}$ whose ijth element is the sum of the products of the corresponding elements of the ith row of A and the jth column of B. When this inner matching of sizes occurs, the matrices A and B are *compatible* for multiplication. Thus,

$$\begin{bmatrix} 3 & 6 & 1 \\ 0 & 5 & 4 \end{bmatrix} \begin{bmatrix} 2 & 4 \\ -3 & 1 \\ 5 & 0 \end{bmatrix} = \begin{bmatrix} (3)(2) + (6)(-3) + (1)(5) & (3)(4) + (6)(1) + (1)(0) \\ (0)(2) + (5)(-3) + (4)(5) & (0)(4) + (5)(1) + (4)(0) \end{bmatrix} = \begin{bmatrix} -7 & 18 \\ 5 & 5 \end{bmatrix}.$$

Note: The product AB is not defined if the number of rows of A is not equal to the number of columns of B; and, in general, $AB \neq BA$.

321. Proceed as in question 320.

$$\begin{bmatrix} 1 & 0 \\ 0 & 1 \end{bmatrix} \begin{bmatrix} 2 & 6 \\ -4 & 7 \end{bmatrix} = \begin{bmatrix} (1)(2) + (0)(-4) & (1)(6) + (0)(7) \\ (0)(2) + (1)(-4) & (0)(6) + (1)(7) \end{bmatrix} = \begin{bmatrix} 2 & 6 \\ -4 & 7 \end{bmatrix}$$

322. Proceed as in question 321.

(A) $\begin{bmatrix} -4 & 5 \\ 1 & 2 \end{bmatrix} \begin{bmatrix} 3 & -1 \\ -6 & 2 \end{bmatrix} = \begin{bmatrix} -42 & 14 \\ -9 & 3 \end{bmatrix}$

(B) $\begin{bmatrix} 3 & -1 \\ -6 & 2 \end{bmatrix} \begin{bmatrix} -4 & 5 \\ 1 & 2 \end{bmatrix} = \begin{bmatrix} -13 & 13 \\ 26 & -26 \end{bmatrix}$

(C) No.

323. The determinant of a square matrix A, denoted det A or $|A|$, is a scalar associated with the matrix and its elements. *Note*: The determinant is not meaningful for non-square matrices.

A second order (2×2) determinant is defined as follows: $\det \begin{bmatrix} a & b \\ c & d \end{bmatrix} = \begin{vmatrix} a & b \\ c & d \end{vmatrix} = ad - bc.$

Thus, $\begin{vmatrix} -4 & 5 \\ 1 & 2 \end{vmatrix} = (-4)(2) - (5)(1) = -8 - 5 = -13.$

324. Proceed as in question 323.

$$\begin{vmatrix} 3 & -1 \\ -6 & 2 \end{vmatrix} = (3)(2) - (-1)(-6) = 6 - 6 = 0$$

325. A third-order (3×3) determinant can be computed using the method of *expanding* by minors about the first row as shown here:

$$\begin{vmatrix} a_1 & a_2 & a_3 \\ b_1 & b_2 & b_3 \\ c_1 & c_2 & c_3 \end{vmatrix} = \begin{vmatrix} a_1 & a_2 & a_3 \\ b_1 & b_2 & b_3 \\ c_1 & c_2 & c_3 \end{vmatrix} = a_1 \begin{vmatrix} b_2 & b_3 \\ c_2 & c_3 \end{vmatrix} - a_2 \begin{vmatrix} b_1 & b_3 \\ c_1 & c_3 \end{vmatrix} + a_3 \begin{vmatrix} b_1 & b_2 \\ c_1 & c_2 \end{vmatrix}$$

Tip: To find the 2×2 determinant, called the *minor*, associated with a_1, mentally cross out the row and column containing a_1 and the remaining 2×2 determinant is the one needed. A similar technique is used to determine the other two minors. *Caution*: In applying this definition, errors are frequently made by forgetting to include the negative sign on the second term of the expansion.

Thus, $\begin{vmatrix} 3 & 4 & 2 \\ -2 & 8 & 2 \\ -2 & 1 & 5 \end{vmatrix} = 3 \begin{vmatrix} 8 & 2 \\ 1 & 5 \end{vmatrix} - 4 \begin{vmatrix} -2 & 2 \\ -2 & 5 \end{vmatrix} + 2 \begin{vmatrix} -2 & 8 \\ -2 & 1 \end{vmatrix} = 3(38) - 4(-6) + 2(14) = 166.$

Note: You can compute the determinant of a square matrix using expansion by minors about any row or column. However, you must multiply the row or column's *ij*th element by $(-1)^{i+j}$ in the expansion.

326. $1 \begin{vmatrix} 1 & 0 \\ 0 & 1 \end{vmatrix} - 0 \begin{vmatrix} 0 & 0 \\ 0 & 1 \end{vmatrix} + 0 \begin{vmatrix} 0 & 1 \\ 0 & 0 \end{vmatrix} = 1(1 - 0) = 1$

327. Not applicable since the matrix is not square.

328. (a) $A^{-1} = \begin{bmatrix} 4 & 1 \\ -3 & -2 \end{bmatrix}^{-1} = \frac{1}{-8+3} \begin{bmatrix} -2 & -1 \\ 3 & 4 \end{bmatrix} = \frac{1}{-5} \begin{bmatrix} -2 & -1 \\ 3 & 4 \end{bmatrix} = \begin{bmatrix} \dfrac{2}{5} & \dfrac{1}{5} \\ -\dfrac{3}{5} & -\dfrac{4}{5} \end{bmatrix}$

(b) $B^{-1} = \begin{bmatrix} -1 & 1 \\ -2 & 4 \end{bmatrix}^{-1} = \frac{1}{-4+2} \begin{bmatrix} 4 & -1 \\ 2 & -1 \end{bmatrix} = \frac{1}{-2} \begin{bmatrix} 4 & -1 \\ 2 & -1 \end{bmatrix} = \begin{bmatrix} -2 & \dfrac{1}{2} \\ -1 & \dfrac{1}{2} \end{bmatrix}$

329. First find AB, and then proceed according to the guidelines in question 328 to find $(AB)^{-1}$.

$$(AB)^{-1} = \left(\begin{bmatrix} 4 & 1 \\ -3 & -2 \end{bmatrix} \begin{bmatrix} -1 & 1 \\ -2 & 4 \end{bmatrix} \right)^{-1} = \left(\begin{bmatrix} -6 & 8 \\ 7 & -11 \end{bmatrix} \right)^{-1} = \frac{1}{10} \begin{bmatrix} -11 & -8 \\ -7 & -6 \end{bmatrix} = \begin{bmatrix} -\dfrac{11}{10} & -\dfrac{4}{5} \\ -\dfrac{7}{10} & -\dfrac{3}{5} \end{bmatrix}$$

330. Using the results from question 328:

$$A^{-1}B^{-1} = \begin{bmatrix} \dfrac{2}{5} & \dfrac{1}{5} \\ -\dfrac{3}{5} & -\dfrac{4}{5} \end{bmatrix} \begin{bmatrix} -2 & \dfrac{1}{2} \\ -1 & \dfrac{1}{2} \end{bmatrix} = \begin{bmatrix} -1 & \dfrac{3}{10} \\ 2 & -\dfrac{7}{10} \end{bmatrix}$$

331. Using the results from question 328:

$$B^{-1}A^{-1} = \begin{bmatrix} -2 & \dfrac{1}{2} \\ -1 & \dfrac{1}{2} \end{bmatrix} \begin{bmatrix} \dfrac{2}{5} & \dfrac{1}{5} \\ -\dfrac{3}{5} & -\dfrac{4}{5} \end{bmatrix} = \begin{bmatrix} -\dfrac{11}{10} & -\dfrac{4}{5} \\ -\dfrac{7}{10} & -\dfrac{3}{5} \end{bmatrix}$$

332. (A) True, based on questions 329 and 331.

     (B) False, based on questions 329 and 330.

333. $A = \begin{bmatrix} 3 & 2 \\ 1 & -2 \end{bmatrix}$, $X = \begin{bmatrix} x_1 \\ x_2 \end{bmatrix}$, and $C = \begin{bmatrix} 4 \\ 3 \end{bmatrix}$

$$\begin{bmatrix} x_1 \\ x_2 \end{bmatrix} = X = A^{-1}C = \begin{bmatrix} 3 & 2 \\ 1 & -2 \end{bmatrix}^{-1} \begin{bmatrix} 4 \\ 3 \end{bmatrix} = \frac{1}{-8} \begin{bmatrix} -2 & -2 \\ -1 & 3 \end{bmatrix} \begin{bmatrix} 4 \\ 3 \end{bmatrix} = -\frac{1}{8} \begin{bmatrix} -14 \\ 5 \end{bmatrix} = \begin{bmatrix} \dfrac{7}{4} \\ -\dfrac{5}{8} \end{bmatrix}$$

Thus, $x_1 = \dfrac{7}{4}$, $x_2 = -\dfrac{5}{8}$.

334. $A = \begin{bmatrix} 5 & -1 \\ 2 & 2 \end{bmatrix}$, $X = \begin{bmatrix} x_1 \\ x_2 \end{bmatrix}$, and $C = \begin{bmatrix} 3 \\ 2 \end{bmatrix}$

$$\begin{bmatrix} x_1 \\ x_2 \end{bmatrix} = X = A^{-1}C = \begin{bmatrix} 5 & -1 \\ 2 & 2 \end{bmatrix}^{-1} \begin{bmatrix} 3 \\ 2 \end{bmatrix} = \frac{1}{12} \begin{bmatrix} 2 & 1 \\ -2 & 5 \end{bmatrix} \begin{bmatrix} 3 \\ 2 \end{bmatrix} = \frac{1}{12} \begin{bmatrix} 8 \\ 4 \end{bmatrix} = \begin{bmatrix} \dfrac{2}{3} \\ \dfrac{1}{3} \end{bmatrix}$$

Thus, $x_1 = \dfrac{2}{3}$, $x_2 = \dfrac{1}{3}$.

335. $D = \begin{vmatrix} 2 & -3 \\ 5 & -2 \end{vmatrix} = 11 \neq 0$

$$x_1 = \frac{\begin{vmatrix} 16 & -3 \\ -4 & -2 \end{vmatrix}}{11} = \frac{-44}{11} = -4, \quad x_2 = \frac{\begin{vmatrix} 2 & 16 \\ 5 & -4 \end{vmatrix}}{11} = \frac{-88}{11} = -8$$

336. $D = \begin{vmatrix} 1 & -2 \\ 3 & 2 \end{vmatrix} = 8 \neq 0$

$$x_1 = \frac{14 + 10}{8} = \frac{24}{8} = 3$$

$$x_2 = \frac{5 - 21}{8} = \frac{-16}{8} = -2$$

337. $D = \begin{vmatrix} 1 & -2 & -3 \\ 2 & 4 & -5 \\ 3 & 7 & -4 \end{vmatrix} = 1(-16 + 35) + 2(-8 + 15) - 3(14 - 12) = 27 \neq 0$

$$x_1 = \frac{\begin{vmatrix} -20 & -2 & -3 \\ 11 & 4 & -5 \\ 33 & 7 & -4 \end{vmatrix}}{27} = \frac{-20(19) + 2(121) - 3(-55)}{27} = \frac{27}{27} = 1$$

$$x_2 = \frac{\begin{vmatrix} 1 & -20 & -3 \\ 2 & 11 & -5 \\ 3 & 33 & -4 \end{vmatrix}}{27} = \frac{1(121) + 20(7) - 3(33)}{27} = \frac{162}{27} = 6$$

$$x_3 = \frac{(1 - 12 + 20)}{3} = 3,$$

obtained by solving the first equation of the system for x_3.

338. $D = \begin{vmatrix} 1 & 2 & -1 \\ 4 & 3 & 2 \\ 9 & 8 & 3 \end{vmatrix} = 1(-7) - 2(-6) - 1(5) = 0$

$$X_1 = \begin{vmatrix} 7 & 2 & -1 \\ 1 & 3 & 2 \\ 4 & 8 & 3 \end{vmatrix} = 7(-7) - 2(-5) - 1(-4) = -35 \neq 0$$

Thus, there is no solution.

339. $D = \begin{vmatrix} 2 & -4 & 7 \\ 3 & 2 & -1 \\ 1 & -10 & 15 \end{vmatrix} = 2(20) + 4(46) + 7(-32) = 0$

$X_1 = \begin{vmatrix} 5 & -4 & 7 \\ 2 & 2 & -1 \\ 8 & -10 & 15 \end{vmatrix} = 5(20) + 4(38) + 7(-36) = 0$

$X_2 = \begin{vmatrix} 2 & 5 & 7 \\ 3 & 2 & -1 \\ 1 & 8 & 15 \end{vmatrix} = 2(38) - 5(46) + 7(22) = 0$

$X_3 = \begin{vmatrix} 2 & -4 & 5 \\ 3 & 2 & 2 \\ 1 & -10 & 8 \end{vmatrix} = 2(36) + 4(22) + 5(-32) = 0$

Therefore, there are infinitely many solutions.

To express the solutions in a general form, solve the 2×2 system $\begin{aligned} 2x_1 - 4x_2 &= 5 - 7x_3 \\ 3x_1 + 2x_2 &= 2 + x_3 \end{aligned}$ for x_1 and x_2, while manipulating x_3 as you would a constant. (*Note*: This system was carefully selected so that the determinant of the 2×2 coefficient matrix would be nonzero; that is, $D \neq 0$.) Now, proceeding as for 2×2 systems, you have

$$x_1 = \frac{\begin{vmatrix} 5 - 7x_3 & -4 \\ 2 + x_3 & 2 \end{vmatrix}}{16} = \frac{10 - 14x_3 + 8 + 4x_3}{16} = \frac{18 - 10x_3}{16} = \frac{9 - 5x_3}{8}$$

$$x_2 = \frac{\begin{vmatrix} 2 & 5 - 7x_3 \\ 3 & 2 + x_3 \end{vmatrix}}{16} = \frac{4 + 2x_3 - 15 + 21x_3}{16} = \frac{23x_3 - 11}{16}$$

The system solution is $\left(\dfrac{9 - 5x_3}{8}, \dfrac{23x_3 - 11}{16}, x_3 \right)$, where x_3 is a "free" variable, meaning that it is allowed to assume any value and the values of the other variables, x_1 and x_2, are dependent on x_3's value. Thus, there are infinitely many solutions. One easily determined particular solution is $\left(\dfrac{9}{8}, -\dfrac{11}{16}, 0 \right)$, which is obtained by letting $x_3 = 0$.

340. Systematically perform elementary row operations on the augmented matrix
$\begin{bmatrix} 3 & 2 & 4 \\ 1 & -2 & 3 \end{bmatrix}$.

Interchange the two rows to obtain $\begin{bmatrix} 1 & -2 & 3 \\ 3 & 2 & 4 \end{bmatrix}$.

Multiply the first row by -3 and add to the second row to obtain $\begin{bmatrix} 1 & -2 & 3 \\ 0 & 8 & -5 \end{bmatrix}$.

Multiply the second row by $\dfrac{1}{4}$ and add to the first row $\begin{bmatrix} 1 & 0 & \frac{7}{4} \\ 0 & 8 & -5 \end{bmatrix}$.

Multiply the second row by $\dfrac{1}{8}$ to obtain $\begin{bmatrix} 1 & 0 & \frac{7}{4} \\ 0 & 1 & -\frac{5}{8} \end{bmatrix}$.

Thus, the system has solution $x_1 = \dfrac{7}{4}$, $x_2 = -\dfrac{5}{8}$.

341. Systematically perform elementary row operations on the augmented matrix
$\begin{bmatrix} 1 & -3 & 1 & 2 \\ 2 & -1 & -2 & 1 \\ 3 & 2 & -1 & 5 \end{bmatrix}$.

Multiply the first row by -2 and add to the second row and at the same time multiply the first row by -3 and add to the third row to obtain $\begin{bmatrix} 1 & -3 & 1 & 2 \\ 0 & 5 & -4 & -3 \\ 0 & 11 & -4 & -1 \end{bmatrix}$.

Multiply the second row by $\dfrac{1}{5}$ to obtain $\begin{bmatrix} 1 & -3 & 1 & 2 \\ 0 & 1 & -0.8 & -0.6 \\ 0 & 11 & -4 & -1 \end{bmatrix}$.

Multiply the second row by 3 and add to the first row and at the same time multiply the second row by -11 and add to the third row to obtain $\begin{bmatrix} 1 & 0 & -1.4 & 0.2 \\ 0 & 1 & -0.8 & -0.6 \\ 0 & 0 & 4.8 & 5.6 \end{bmatrix}$.

Multiply the third row by $\dfrac{1}{4.8} = \dfrac{5}{24}$ to obtain $\begin{bmatrix} 1 & 0 & -1.4 & 0.2 \\ 0 & 1 & -0.8 & -0.6 \\ 0 & 0 & 1 & \frac{7}{6} \end{bmatrix}$.

Multiply the third row by 1.4 and add to the first row and at the same time multiply the

third row by 0.8 and add to the second row to obtain $\begin{bmatrix} 1 & 0 & 0 & \frac{11}{6} \\ 0 & 1 & 0 & \frac{1}{3} \\ 0 & 0 & 1 & \frac{7}{6} \end{bmatrix}$.

Thus, the system has solution $x_1 = \frac{11}{6}, x_2 = \frac{1}{3}, x_3 = \frac{7}{6}$.

342. Proceed as in questions 340 and 341. The symbol ~ is used to mean "is equivalent to."

$$\begin{bmatrix} 2 & -1 & 0 & 0 \\ 0 & 2 & -1 & 0 \\ 1 & 2 & -1 & 3 \end{bmatrix} \sim \begin{bmatrix} 1 & 2 & -1 & 3 \\ 0 & 2 & -1 & 0 \\ 2 & -1 & 0 & 0 \end{bmatrix} \sim \begin{bmatrix} 1 & 2 & -1 & 3 \\ 0 & 2 & -1 & 0 \\ 0 & -5 & 2 & -6 \end{bmatrix} \sim \begin{bmatrix} 1 & 2 & -1 & 3 \\ 0 & 1 & -0.5 & 0 \\ 0 & -5 & 2 & -6 \end{bmatrix}$$

$$\sim \begin{bmatrix} 1 & 0 & 0 & 3 \\ 0 & 1 & -0.5 & 0 \\ 0 & 0 & -0.5 & -6 \end{bmatrix} \sim \begin{bmatrix} 1 & 0 & 0 & 3 \\ 0 & 1 & -0.5 & 0 \\ 0 & 0 & 1 & 12 \end{bmatrix} \sim \begin{bmatrix} 1 & 0 & 0 & 3 \\ 0 & 1 & 0 & 6 \\ 0 & 0 & 1 & 12 \end{bmatrix}$$

The system has solution $x_1 = 3, x_2 = 6, x_3 = 12$.

Chapter 8: Sequences, Series, and Mathematical Induction

343. Arithmetic sequences have the form $a_1, a_1 + d, a_1 + 2d, a_1 + 3d, a_1 + 4d, \ldots, a_1 + (n-1)d, \ldots$, where a_1 is the first term and $d = a_{n+1} - a_n$ is the common difference between successive terms. The closed formula for the nth term is $a_n = a_1 + (n-1)d$, $n = 1, 2, 3, \ldots$. Inspection of the arithmetic sequence 2, 5, 8, 11, 14, ... reveals that $a_1 = 2$ and $d = 3$. Thus, the closed formula for its nth term is $a_n = a_1 + (n-1)d = 2 + (n-1)3 = 3n - 1$.

344. Inspection of the arithmetic sequence 6, 4, 2, 0, ... reveals that $a_1 = 6$ and $d = -2$. Thus, the closed formula for its nth term is $a_n = a_1 + (n-1)d = 6 + (n-1)(-2) = -2n + 8$.

345. Inspection of the arithmetic sequence $\frac{7}{3}, \frac{16}{3}, \frac{25}{3}, \frac{34}{3}, \ldots$ reveals that $a_1 = \frac{7}{3}$ and $d = \frac{9}{3}$. Thus, the closed formula for its nth term is $a_n = a_1 + (n-1)d = \frac{7}{3} + (n-1)\frac{9}{3} = \frac{9}{3}n - \frac{2}{3}$.

346. The recursive formula for an arithmetic sequence is $a_1 = a_1$ and $a_{n+1} = a_n + d$, where a_1 is the first term and $n = 1, 2, 3, \ldots$. From question 343, $a_1 = 2$ and $d = 3$, so the recursive formula for 2, 5, 8, 11, 14, ... is $a_1 = 2$ and $a_{n+1} = a_n + 3$.

347. From question 344, $a_1 = 6$ and $d = -2$, so the recursive formula for 6, 4, 2, 0, ... is $a_1 = 6$ and $a_{n+1} = a_n + (-2)$.

348. Geometric sequences have the form $a_1, a_1 r, a_1 r^2, a_1 r^3, \ldots, a_1 r^{n-1}, \ldots$, where a_1 is the first term and $r = \dfrac{a_{n+1}}{a_n}$ is the common ratio between successive terms. The closed formula for the nth term is $a_n = a_1 r^{n-1}$, $n = 1, 2, 3, \ldots$. Because the sequence 2, 6, 18, 54, ... is geometric with $a_1 = 2$ and $a_2 = 6$, the common ratio $r = \dfrac{6}{2} = 3$. Thus, the closed formula of the sequence is $a_n = a_1 r^{n-1} = 2(3)^{n-1}$.

349. Because the sequence $\dfrac{1}{3}, \dfrac{2}{15}, \dfrac{4}{75}, \dfrac{8}{375}, \ldots$ is geometric with $a_1 = \dfrac{1}{3}$ and $a_2 = \dfrac{2}{15}$, the common ratio $r = \dfrac{\frac{2}{15}}{\frac{1}{3}} = \dfrac{2}{5}$. Thus, the closed formula of the sequence is $a_n = a_1 r^{n-1} = \dfrac{1}{3}\left(\dfrac{2}{5}\right)^{n-1}$.

350. Because the sequence $3, -6, 12, -24, \ldots$ is geometric with $a_1 = 3$ and $a_2 = -6$, the common ratio $r = \dfrac{-6}{3} = -2$. Thus, the closed formula of the sequence is $a_n = a_1 r^{n-1} = 3(-2)^{n-1}$.

351. The recursive formula for a geometric sequence is $a_1 = a_1$ and $a_{n+1} = r a_n$, $n = 1, 2, 3, \ldots$. From question 348, $a_1 = 2$ and $r = 3$, so the recursive formula for 2, 6, 18, 54, ... is $a_1 = 2$ and $a_{n+1} = 3 a_n$.

352. From question 349, $a_1 = \dfrac{1}{3}$ and $r = \dfrac{2}{5}$, so the recursive formula for $\dfrac{1}{3}, \dfrac{2}{15}, \dfrac{4}{75}, \dfrac{8}{375}, \ldots$ is $a_1 = \dfrac{1}{3}$ and $a_{n+1} = \dfrac{2}{5} a_n$.

353. (A) $a_1 = 3(1) + 2$, $a_2 = 3(2) + 2$, $a_3 = 3(3) + 2$, $a_4 = 3(4) + 2 : 5, 8, 11, 14$

(B) $a_{17} = 3(17) + 2 = 53$

354. (A) $a_1 = \dfrac{1}{2}(1) - 1$, $a_2 = \dfrac{1}{2}(2) - 1$, $a_3 = \dfrac{1}{2}(3) - 1$, $a_4 = \dfrac{1}{2}(4) - 1 : -\dfrac{1}{2}, 0, \dfrac{1}{2}, 1$

(B) $a_{19} = \dfrac{1}{2}(19) - 1 = \dfrac{17}{2}$

355. (A) $a_1 = -2(1) - 1$, $a_2 = -2(2) - 1$, $a_3 = -2(3) - 1$, $a_4 = -2(4) - 1 : -3, -5, -7, -9$

(B) $a_{100} = -2(100) - 1 = -201$

356. (A) $a_1 = 2\left(\dfrac{1}{3}\right)^{1-1}$, $a_2 = 2\left(\dfrac{1}{3}\right)^{2-1}$, $a_3 = 2\left(\dfrac{1}{3}\right)^{3-1}$, $a_4 = 2\left(\dfrac{1}{3}\right)^{4-1}$: 2, $\dfrac{2}{3}$, $\dfrac{2}{9}$, $\dfrac{2}{27}$

(B) $a_7 = 2\left(\dfrac{1}{3}\right)^6 = \dfrac{2}{729}$

357. (A) $a_1 = 3\left(-\dfrac{1}{2}\right)^{1-1}$, $a_2 = 3\left(-\dfrac{1}{2}\right)^{2-1}$, $a_3 = 3\left(-\dfrac{1}{2}\right)^{3-1}$, $a_4 = 3\left(-\dfrac{1}{2}\right)^{4-1}$:

3, $-\dfrac{3}{2}$, $\dfrac{3}{4}$, $-\dfrac{3}{8}$

(B) $a_7 = 3\left(-\dfrac{1}{2}\right)^6 = \dfrac{3}{64}$

358. Use the common difference $d = a_{n+1} - a_n$ or common ratio $r = \dfrac{a_{n+1}}{a_n}$ formulas to check.

$$0.5 - 1 = -0.5; \ 0.25 - 0.5 = -0.25$$

These differences are not the same, so the sequence is not arithmetic.

$$\dfrac{0.5}{1} = 0.5; \ \dfrac{0.25}{0.5} = 0.5; \ \dfrac{0.125}{0.25} = 0.5$$

The ratios are the same so the sequence is geometric.

359. Use the common difference $d = a_{n+1} - a_n$ or common ratio $r = \dfrac{a_{n+1}}{a_n}$ formulas to check.

$$\dfrac{3}{4} - \dfrac{1}{4} = \dfrac{1}{2}; \ \dfrac{3}{8} - \dfrac{3}{4} = -\dfrac{3}{8}$$

These differences are not the same, so the sequence is not arithmetic.

$$\dfrac{3}{4} \div \dfrac{1}{4} = 3; \ \dfrac{3}{8} \div \dfrac{3}{4} = \dfrac{1}{2}$$

The ratios are not the same so the sequence is not geometric.

Thus, the sequence $\dfrac{1}{4}, \dfrac{3}{4}, \dfrac{3}{8}, 1, \ldots$ is neither arithmetic nor geometric.

360. The sequence 2, 2, 2, 2, … is arithmetic with $d = 0$ and geometric with $r = 1$.

361. These are terms of an arithmetic sequence whose closed formula is $a_n = 2 + (n-1)5 = 5n - 3$. The number 147 must satisfy this formula. Therefore, $147 = 5n - 3$. Solving this equation for n yields $n = \dfrac{147 + 3}{5} = \dfrac{150}{5} = 30$. Thus, there are 30 terms in the list.

362. These are terms of a geometric sequence whose closed formula is $a_n = 2(2)^{n-1} = 2^n$. The number 4,096 must satisfy this formula. Therefore, $4{,}096 = 2^n$. Solving using logarithms,

$$2^n = 4{,}096$$
$$\log_2(2^n) = \log_2(4{,}096)$$
$$n\log_2(2) = \log_2(4{,}096)$$
$$n = \log_2(4{,}096) = \frac{\ln 4{,}096}{\ln 2} = 12$$

Thus, there are 12 terms included in the list.

(*Note*: Trial-and-error guessing also leads to the solution $n = 12$.)

363. $3 + 8 + \displaystyle\sum_{k=3}^{15}(5k - 2)$

364. Make the following symbol substitution: $k = j - 1$. Then when $k = 1, j = 2$, and when $k = n, j = n + 1$. So you have $\displaystyle\sum_{k=1}^{n}(5k - 2) = \sum_{j=2}^{n+1}(5j - 7)$.

365. $\displaystyle\sum_{k=1}^{15}(5k - 2) = 5\sum_{k=1}^{15}k - \sum_{k=1}^{15}2 = 5\sum_{k=1}^{15}k - 15(2) = 5\sum_{k=1}^{15}k - 30$

366. $\displaystyle\sum_{k=1}^{n}(5k - 2) = 3 + 8 + 13 + 18 + \ldots + (5n - 2)$

367. $\displaystyle\sum_{k=1}^{6}(5k - 2) = 3 + 8 + 13 + 18 + 23 + 28 = 93$

368. Applying $S_n = \displaystyle\sum_{k=1}^{n}a_1 r^{k-1} = \frac{a_1(1 - r^n)}{1 - r}$, where $a_1 = 1$, $r = \dfrac{1}{2}$, and $n = 7$ yields

$$S_7 = \sum_{k=1}^{7}\left(\frac{1}{2}\right)^{k-1} = \frac{1\left(1 - \left(\frac{1}{2}\right)^7\right)}{1 - \frac{1}{2}} = 2\left(1 - \frac{1}{128}\right) = \frac{127}{64}.$$

369. Applying $S_n = \displaystyle\sum_{k=1}^{n}[a_1 + (k - 1)d] = \frac{n(a_1 + a_n)}{2}$, where $a_1 = 3$, $d = 2$, and $n = 12$ yields

$$S_{12} = \sum_{k=1}^{12}[3 + (k - 1)2] = \frac{12[3 + 3 + (12 - 1)2]}{2} = 6(3 + 25) = 168.$$

370. This sequence is arithmetic with $a_1 = 4$, $a_8 = 25$, and $n = 8$. Applying

$$S_n = \sum_{k=1}^{n} [a_1 + (k-1)d] = \frac{n(a_1 + a_n)}{2} \text{ yields } 4 + 7 + 10 + 13 + 16 + 19 + 22 + 25 =$$

$$S_8 = \frac{8(4 + 25)}{2} = 116.$$

371. This sequence is geometric with $a_1 = 1$, $r = 2$, and $n = 8$. Applying $\sum_{k=1}^{n} a_1 r^{k-1} = \frac{a_1(1 - r^n)}{1 - r}$

yields $1 + 2 + 4 + 8 + 16 + 32 + 64 + 128 = \frac{1(1 - 2^8)}{1 - 2} = 2^8 - 1 = 255.$

372. This is an arithmetic sequence but you need to know how many terms are included in order to sum it using the formula. Since 477 is a term of the sequence, it must satisfy the term formula, $477 = 3 + (n - 1)2 = 2n + 1$. Solving this for n gives $n = 238$. Now applying

$$S_n = \sum_{k=1}^{n} [a_1 + (k+1)d] = \frac{n(a_1 + a_n)}{2} \text{ yields } S_{238} = \frac{238(3 + 477)}{2} = 57{,}120.$$

373. For $n = 1$,

$1(1 + 1) = 2$ so S_1 is true.

If $2 + 4 + \ldots + 2k = k(k + 1)$, then

$2 + 4 + \ldots + 2k + 2(k + 1) = k(k + 1) + 2(k + 1) = (k + 1)(k + 2)$.

Thus, S_{k+1} is true. Therefore, S_n is true for all n.

374. For $n = 1$,

$1(2 + 3) = 5$ so S_1 is true.

If $5 + 9 + \ldots + (4k + 1) = k(2k + 3)$, then

$5 + 9 + \ldots + (4k + 1) + [4(k + 1) + 1]$

$= k(2k + 3) + 4(k + 1) + 1$

$= 2k^2 + 7k + 5$

$= (k + 1)(2k + 5)$

$= (k + 1)[2(k + 1) + 3]$.

Thus, S_{k+1} is true. Therefore, S_n is true for all n.

375. For $n = 1$,

$2^{1+1} - 2 = 2$ so S_1 is true.

If $2 + 4 + \ldots + 2^k = 2^{k+1} - 2$, then

$2 + 4 + \ldots + 2^k + 2^{k+1}$

$= 2^{k+1} - 2 + 2^{k+1}$

$= 2^{k+1}(2) - 2$

$= 2^{k+2} - 2$.

Thus, S_{k+1} is true. Therefore, S_n is true for all n.

376. For $n = 1$,

$$\frac{1}{2 \cdot 1 + 1} = \frac{1}{3} \text{ so } S_1 \text{ is true.}$$

If $\dfrac{1}{3} + \dfrac{1}{15} + \dfrac{1}{35} + \ldots + \dfrac{1}{(2k-1)(2k+1)} = \dfrac{k}{2k+1}$, then

$$\frac{1}{3} + \frac{1}{15} + \frac{1}{35} + \ldots + \frac{1}{(2k-1)(2k+1)} + \frac{1}{(2k+1)(2k+3)}$$

$$= \frac{k}{2k+1} + \frac{1}{(2k+1)(2k+3)}$$

$$= \frac{1}{2k+1} \left[\frac{2k^2 + 3k + 1}{2k+3} \right]$$

$$= \frac{(2k+1)(k+1)}{(2k+1)(2k+3)}$$

$$= \frac{k+1}{2k+3}.$$

Thus, S_{k+1} is true. Therefore, S_n is true for all n.

377. For $n = 1$,

$$\frac{3(3^1 - 1)}{2} = \frac{6}{2} = 3 \text{ so } S_1 \text{ is true.}$$

If $3 + 9 + 27 + \ldots + 3^k = \dfrac{3(3^k - 1)}{2}$, then

$$= 3 + 9 + 27 + \ldots + 3^k + 3^{k+1}$$

$$= \frac{3(3^k - 1)}{2} + 3^{k+1}$$

$$= \frac{3^{k+1} - 3 + 2(3^{k+1})}{2}$$

$$= \frac{3^{k+1}(3) - 3}{2}$$

$$= \frac{3(3^{k+1} - 1)}{2}.$$

Thus, S_{k+1} is true. Therefore, S_n is true for all n.

378. For $n = 1$,

$2^{1+1} = 4 > 2 \cdot 1 + 1 = 3$ so S_1 is true.

If $2^{k+1} \geq 2k + 1$, then

$2^{k+2} = 2(2^{k+1}) \geq 2(2k + 1) = 4k + 2$

$\geq 2k + 2k + 2 \geq 2(k + 1) + 2k \geq 2(k + 1) + 1.$

Thus, S_{k+1} is true. Therefore, S_n is true for all n.

379. For $n = 1$,

$\dfrac{1}{1+1} = \dfrac{1}{2}$ so S_1 is true.

If $\dfrac{1}{2} + \dfrac{1}{6} + \dfrac{1}{12} + \ldots + \dfrac{1}{k(k+1)} = \dfrac{k}{k+1}$, then

$\dfrac{1}{2} + \dfrac{1}{6} + \dfrac{1}{12} + \ldots + \dfrac{1}{k(k+1)} + \dfrac{1}{(k+1)(k+2)}$

$= \dfrac{k}{(k+1)} + \dfrac{1}{(k+1)(k+2)}$

$= \dfrac{1}{(k+1)} \left(\dfrac{k^2 + 2k + 1}{k+2} \right)$

$= \dfrac{(k+1)^2}{(k+1)(k+2)}$

$= \dfrac{k+1}{k+2}.$

Thus, S_{k+1} is true. Therefore, S_n is true for all n.

380. For $n = 1$,

$\dfrac{a(1 - r^1)}{1 - r} = a$ so S_1 is true.

If $a + ar + ar^2 + ar^3 + \ldots + ar^{k-1} = \dfrac{a(1 - r^k)}{1 - r}$, then

$a + ar + ar^2 + ar^3 + \ldots + ar^{k-1} + ar^k$

$= \dfrac{a(1 - r^k)}{1 - r} + ar^k$

$= \dfrac{a - ar^k + ar^k - ar^{k+1}}{1 - r}$

$= \dfrac{a(1 - r^{k+1})}{1 - r}.$

Thus, S_{k+1} is true. Therefore, S_n is true for all n.

381. For $n = 1$,

$\dfrac{1(1+r)^2}{4} = 1$ so S_1 is true.

If $1 + 8 + 27 + 64 + \ldots + k^3 = \dfrac{k^2(k+1)^2}{4}$, then

$1 + 8 + 27 + 64 + \ldots + k^3 + (k+1)^3$

$= \dfrac{k^2(k+1)^2}{4} + (k+1)^3$

$= (k+1)^2\left(\dfrac{k^2 + 4k + 4}{4}\right)$

$= \dfrac{(k+1)^2(k+2)^2}{4}.$

Thus, S_{k+1} is true. Therefore, S_n is true for all n.

382. For $n = 1$,

$4^1 - 1 = 3 \geq 3(4^0) = 3$ so S_1 is true.

If $4^k - 1 \geq 3(4^{k-1})$, then

$4(4^k - 1) \geq 4(3(4^{k-1}))$

$4^{k+1} - 4 \geq 3(4^k)$

$4^{k+1} - 1 \geq 3(4^k) + 3 \geq 3(4^k).$

Thus, S_{k+1} is true. Therefore, S_n is true for all n.

Chapter 9: Trigonometric Functions

383. $45° = 45\left(\dfrac{\pi}{180}\text{ radians}\right) = \dfrac{\pi}{4}\text{ radians}$

384. $1\text{ radian} = 1\left(\dfrac{180°}{\pi}\right) \approx 57.296°$

385. $60° = 60\left(\dfrac{\pi}{180}\text{ radians}\right) = \dfrac{\pi}{3}\text{ radians}$

386. $\dfrac{\pi}{6}\text{ radians} = \dfrac{\pi}{6}\left(\dfrac{180°}{\pi}\right) = 30°$

387. $127° = 127\left(\dfrac{\pi}{180}\text{ radians}\right) \approx 2.217\text{ radians}$

388. 25 radians $= 25\left(\dfrac{180°}{\pi}\right) \approx 1432.39°$

389. $135° = 135\left(\dfrac{\pi}{180} \text{ radians}\right) = \dfrac{3\pi}{4}$ radians

390. $0° = 0\left(\dfrac{\pi}{180} \text{ radians}\right) = 0$ radians

391. As shown in this problem, fractions of a degree can be expressed in minutes (')
and seconds ("). Recall that 1 minute $= \dfrac{1}{60}°$ and 1 second $= \dfrac{1}{60}' = \dfrac{1}{3600}°$.

$28° \, 43' 25'' = \left[28 + 43\left(\dfrac{1}{60}\right) + 25\left(\dfrac{1}{3600}\right)\right]° = 28.72°$

392. $57.5692° = 57° + (.5692)60' = 57° + 34.152' = 57° + 34' + (.152)60'' = 57° \, 34' \, 9''$

393. Using the $30° - 60° - 90°$ triangle, $\cos 60° = \dfrac{\text{side adjacent}}{\text{hypotenuse}} = \dfrac{1}{2}$.

394. Using the $30° - 60° - 90°$ triangle, $\sin 30° = \dfrac{\text{side opposite}}{\text{hypotenuse}} = \dfrac{1}{2}$.

395. Using the $45° - 45° - 90°$ triangle, $\tan 45° = \dfrac{\text{side opposite}}{\text{side adjacent}} = \dfrac{1}{1} = 1$.

396. Using the unit circle, $\cos \dfrac{3\pi}{2} = \dfrac{0}{1} = 0$.

397. Using the $30° - 60° - 90°$ triangle, $\tan 60° = \dfrac{\text{side opposite}}{\text{side adjacent}} = \dfrac{\sqrt{3}}{1} = \sqrt{3}$.

398. Using the unit circle, $\sin \dfrac{3\pi}{2} = \dfrac{-1}{1} = -1$.

399. Using the unit circle, $\cos \pi = \dfrac{-1}{1} = -1$.

400. Using the $45° - 45° - 90°$ triangle, $\sin 45° = \dfrac{\text{side opposite}}{\text{hypotenuse}} = \dfrac{1}{\sqrt{2}} = \dfrac{\sqrt{2}}{2}$.

401. Using the $45° - 45° - 90°$ triangle, $\cos 45° = \dfrac{\text{side adjacent}}{\text{hypotenuse}} = \dfrac{1}{\sqrt{2}} = \dfrac{\sqrt{2}}{2}$.

402. Using the unit circle, $\dfrac{\pi}{2} = \dfrac{y}{x} = \dfrac{1}{0}$ undefined.

403. The reference angle for 135° is 45°. The cosine is negative in quadrant II so, $\cos 135° = -\dfrac{\sqrt{2}}{2}$.

404. The reference angle for 210° is 30°. The tangent is positive in quadrant III so, $\tan 210° = \dfrac{1}{\sqrt{3}} = \dfrac{\sqrt{3}}{3}$.

405. Using the unit circle, $\cos 2\pi = \dfrac{1}{1} = 1$.

406. Using the unit circle, $\tan \pi = \dfrac{0}{-1} = 0$.

407. Using the unit circle, $\sin \pi = \dfrac{0}{1} = 0$.

408. $\text{period} = \dfrac{2\pi}{|B|} = \dfrac{2\pi}{1} = 2\pi$

409. $\text{period} = \dfrac{2\pi}{|B|} = \dfrac{2\pi}{3}$

410. $\text{period} = \dfrac{2\pi}{|B|} = \dfrac{2\pi}{5}$

411. $\text{period} = \dfrac{2\pi}{|B|} = \dfrac{2\pi}{\dfrac{1}{2}} = 4\pi$

412. $\text{period} = \dfrac{2\pi}{|B|} = \dfrac{2\pi}{1} = 2\pi$

413. (a) phase shift: 2π units left; (b) vertical shift: 4 units up

414. $y = 3\cos\left(2x - \dfrac{\pi}{3}\right) = 3\cos\left(2\left(x - \dfrac{\pi}{6}\right)\right)$

(a) phase shift: $\dfrac{\pi}{6}$ units right; (b) no vertical shift

415. $y = 2\sin\left(3x + \dfrac{\pi}{5}\right) - 1 = 2\sin\left(3\left(x + \dfrac{\pi}{15}\right)\right) - 1$

(a) phase shift: $\dfrac{\pi}{15}$ units left; (b) vertical shift: 1 unit down

416. $y = -2\cos\left(\dfrac{1}{3}x - \pi\right) = -2\cos\left(\dfrac{1}{3}(x - 3\pi)\right)$

(a) phase shift: 3π units right; (b) no vertical shift

417. $y = 4\sin\left(2x - \sqrt{2}\right) + \dfrac{3}{5} = 4\sin\left(2\left(x - \dfrac{\sqrt{2}}{2}\right)\right) + \dfrac{3}{5}$

(a) phase shift: $\dfrac{\sqrt{2}}{2}$ units right; (b) vertical shift: $\dfrac{3}{5}$ units up

418. $A = 3$

419. $A = 4$

420. $A = \dfrac{1}{2}$

421. $A = \pi$

422. $A = 12$

423. (a) period $= \dfrac{\pi}{2}$; (b) no phase shift; (c) no vertical shift

424. $y = \tan(-3\theta + 2\pi) - 5 = \tan\left(-3\left(\theta - \dfrac{2\pi}{3}\right)\right) - 5$

(a) period $= \dfrac{\pi}{3}$; (b) phase shift: $\dfrac{2\pi}{3}$ units right; (c) vertical shift: 5 units down

425. $y = 3\tan(\pi x - 2) + 10 = 3\tan\left(\pi\left(x - \dfrac{2}{\pi}\right)\right) + 10$

(a) period $= \dfrac{\pi}{\pi} = 1$; (b) phase shift: $\dfrac{2}{\pi}$ units right; (c) vertical shift: 10 units up

426. $y = 7 \tan\left(\dfrac{3}{\pi}x + 5\right) = 7 \tan\left(\dfrac{3}{\pi}\left(x + \dfrac{5\pi}{3}\right)\right)$

(a) period $= \dfrac{\pi}{\dfrac{3}{\pi}} = \dfrac{\pi^2}{3}$; (b) phase shift: $\dfrac{5\pi}{3}$ units left; (c) no vertical shift

427. $y = \cos(3x - 6) - 2 = \cos(3(x - 2)) - 2$

(a) period $= \dfrac{2\pi}{3}$; (b) phase shift: 2 units right; (c) vertical shift: 2 units down

428. The solutions are $x = -2\pi, -\pi, 0, \pi, 2\pi$ because in the interval $[-2\pi, 2\pi]$ the graph of $y = \sin x$ intersects the x-axis at these values for x.

429. The solutions are $x = -\dfrac{3\pi}{2}, \dfrac{-\pi}{2}, \dfrac{\pi}{2}, \dfrac{3\pi}{2}$ because in the interval $[-2\pi, 2\pi]$ the graph of $y = \cos x$ intersects the x-axis at these values for x.

430. $y = 3 \sin(2x - \pi) = 3 \sin 2\left(x - \dfrac{\pi}{2}\right)$; amplitude $= 3$, period $= \dfrac{2\pi}{2} = \pi$, phase shift: $\dfrac{\pi}{2}$ right, no vertical shift

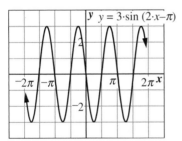

431. $y = -2\cos(-3x - 6\pi) = -2\cos(-3(x + 2\pi))$; reflection about the x-axis, amplitude $= 2$, period $= \dfrac{2\pi}{3}$, phase shift: 2π left, no vertical shift

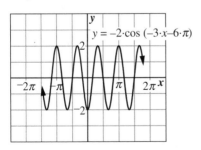

432. $y = \dfrac{1}{2}\tan(3x - 6) = \dfrac{1}{2}\tan(3(x - 2))$; period $= \dfrac{\pi}{3}$, phase shift: 2 units right, no vertical shift

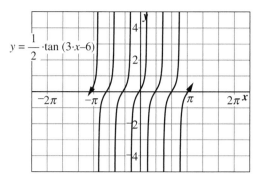

Chapter 10: Analytic Trigonometry

433. In working with identities, you generally start with the side of the identity that looks more complicated. Then use your knowledge of well-known identities and your algebra skills until that side looks exactly like the other side. *Note*: For convenience LS and RS will be used to denote the left side and right side of an identity, respectively.

$$\sin x \cot x = \sin x \frac{\cos x}{\sin x}$$

Starting on the LS, use $\cot x = \dfrac{\cos x}{\sin x}$.

$$= \cos x$$

Simplify. The LS is the same as the RS, so the identity is proven.

434. $\sin x(\csc x - \sin x) = \sin x \csc x - \sin^2 x$

Starting on the LS, remove parentheses.

$$= \sin x \frac{1}{\sin x} - \sin^2 x$$

Use $\csc = \dfrac{1}{\sin x}$.

$$= 1 - \sin^2 x$$

Simplify.

$$= \cos^2 x$$

Use $1 - \sin^2 x = \cos^2 x$. The LS is the same as the RS, so the identity is proven.

435. $(1 + \sin x)(1 + \sin(-x)) = (1 + \sin x)(1 - \sin x)$

Starting on the LS, use $\sin(-x) = -\sin x$.

$$= 1 - \sin^2 x$$

Multiply, recalling that $(x + y)(x - y) = x^2 - y^2$.

$$= \cos^2 x$$

Use $1 - \sin^2 x = \cos^2 x$. The LS is the same as the RS, so the identity is proven.

436. $\tan^2 x \csc^2 x - \tan^2 x = \tan^2 x (\csc^2 x - 1)$ Factor the LS.

$$= \tan^2 x \cot^2 x$$ Use $\csc^2 x - 1 = \cot^2 x$.

$$= \tan^2 x \frac{1}{\tan^2 x}$$ Use $\cot x = \frac{1}{\tan x}$.

$$= 1$$ Simplify. The LS is the same as the RS, so the identity is proven.

437. $\tan x (\csc x + \cot x) = \tan x \csc x + \tan x \cot x$ Starting on the LS, remove parentheses.

$$= \frac{\sec x}{\csc x} \csc x + \tan x \frac{1}{\tan x}$$ Use $\tan x = \frac{\sec x}{\csc x}$ and $\cot x = \frac{1}{\tan x}$.

$$= \sec x + 1$$ Simplify. The LS is the same as the RS, so the identity is proven.

438. $\dfrac{\sin x \cos x + \cos x}{\sin x + \sin^2 x} = \dfrac{\cos x (\sin x + 1)}{\sin x (1 + \sin x)}$ Factor the numerator and denominator on the LS.

$$= \frac{\cos x}{\sin x}$$ Simplify.

$$= \cot x$$ Use $\cot x = \dfrac{\cos x}{\sin x}$. The LS is the same as the RS, so the identity is proven.

439. $\dfrac{\cos^2 x}{\sin x} + \dfrac{\sin x}{1} = \dfrac{\cos^2 x + \sin^2 x}{\sin x}$ Combine fractions on the LS.

$$= \frac{1}{\sin x}$$ Use $\sin^2 x + \cos^2 x = 1$.

$$= \csc x$$ Use $\csc x = \dfrac{1}{\sin x}$. The LS is the same as the RS, so the identity is proven.

440. $\dfrac{\sin x + \cos x}{\tan x} = \dfrac{\sin x}{\tan x} + \dfrac{\cos x}{\tan x}$ Starting on the LS, write as the sum of two fractions.

$$= \frac{\sin x}{\left(\dfrac{\sin x}{\cos x}\right)} + \frac{\cos x}{\left(\dfrac{\sin x}{\cos x}\right)}$$ Use $\tan x = \dfrac{\sin x}{\cos x}$.

$$= \cos x + \frac{\cos^2 x}{\sin x}$$ Simplify. The LS is the same as the RS, so the identity is proven.

441. $(\sin x + \cos x)^2 = \sin^2 x + 2\sin x \cos x + \cos^2 x$ Starting on the LS, multiply, recalling that $(x + y)^2 = x^2 + 2xy + y^2$.

$$= (\sin^2 x + \cos^2 x) + 2\sin x \cos x$$ Regroup.

$$= 1 + 2\sin x \cos x$$ Use $\sin^2 x + \cos^2 x = 1$. The LS is the same as the RS, so the identity is proven.

442. $\sec^4 x - \tan^4 x = (\sec^2 x - \tan^2 x)(\sec^2 x + \tan^2 x)$ Factor the LS.

$$= (1)(\sec^2 x + \tan^2 x)$$ Use $\sec^2 x - \tan^2 x = 1$.

$$= \sec^2 x + \tan^2 x$$ Simplify. The LS is the same as the RS, so the identity is proven.

443. $\sin(x + 2\pi) = \sin x \cos(2\pi) + \cos x \sin(2\pi)$

$$= \sin x(1) + \cos x(0)$$
$$= \sin x$$

444. $\cos(x + 2\pi) = \cos x \cos(2\pi) - \sin x \sin(2\pi)$

$$= \cos x(1) - \sin x(0)$$
$$= \cos x$$

445. $\tan(x + \pi) = \dfrac{\tan x + \tan \pi}{1 - \tan x \tan \pi}$

$$= \dfrac{\tan x + 0}{1 - \tan x(0)}$$

$$= \dfrac{\tan x}{1 - 0}$$

$$= \tan x$$

446. $\cos\left(x + \dfrac{\pi}{4}\right) = \cos x \cos\left(\dfrac{\pi}{4}\right) - \sin x \sin\dfrac{\pi}{4}$

$$= \dfrac{\sqrt{2}}{2}\cos x - \dfrac{\sqrt{2}}{2}\sin x$$

$$= \dfrac{\sqrt{2}}{2}(\cos x - \sin x)$$

447. $\sin(\alpha + \beta)\sin(\alpha - \beta) = (\sin\alpha\cos\beta + \cos\alpha\sin\beta)(\sin\alpha\cos\beta - \cos\alpha\sin\beta)$

$= \sin^2\alpha\cos^2\beta - \cos^2\alpha\sin^2\beta$

$= \sin^2\alpha\cos^2\beta + \sin^2\alpha\sin^2\beta - \sin^2\alpha\sin^2\beta - \cos^2\alpha\sin^2\beta$ (Notice the strategy of

$= \sin^2\alpha(\cos^2\beta + \sin^2\beta) - \sin^2\beta(\cos^2\alpha + \sin^2\alpha)$ adding and subtracting

$= \sin^2\alpha - \sin^2\beta$ the same quantity.)

448. $\tan\left(x + \dfrac{\pi}{4}\right) = \dfrac{\tan x + \tan\dfrac{\pi}{4}}{1 - \tan x \tan\dfrac{\pi}{4}}$

$= \dfrac{\tan x + (1)}{1 - \tan x(1)}$

$= \dfrac{\tan x + 1}{1 - \tan x}$

$= \dfrac{1 + \tan x}{1 - \tan x}$

449. $\sin(2\theta) = \sin(\theta + \theta)$

$= \sin\theta\cos\theta + \cos\theta\sin\theta$

$= 2\sin\theta\cos\theta$

450. $\cos(2\theta) = \cos(\theta + \theta)$

$= \cos\theta\cos\theta - \sin\theta\sin\theta$

$= \cos^2\theta - \sin^2\theta$

$= \cos^2\theta - (1 - \cos^2\theta)$

$= 2\cos^2\theta - 1$

451. $\sin(4x) = \sin(2(2x))$

$= 2\sin(2x)\cos(2x)$ By question 449, $\sin(2\theta) = 2\sin\theta\cos\theta$, so let $\theta = 2x$.

$= 2(2\sin x\cos x)(2\cos^2 x - 1)$ By question 450, $\cos(2\theta) = 2\cos^2\theta - 1$,

$= 2(2\sin x\cos x)(2(1 - 2\sin^2 x) - 1)$ so let $\theta = x$.

$= 2(2\sin x\cos x)(1 - 2\sin^2 x)$

$= 4\sin x\cos x(1 - 2\sin^2 x)$

452. $\dfrac{\cos(2x)}{\sin^2 x} = \dfrac{2\cos^2 x - 1}{\sin^2 x}$

$\qquad = 2\left(\dfrac{\cos^2 x}{\sin^2 x}\right) - \dfrac{1}{\sin^2 x}$

$\qquad = 2\cot^2 x - \csc^2 x$

$\qquad = 2\cot^2 x - (1 + \cot^2 \theta)$

$\qquad = \cot^2 x - 1$

By question 450, $\cos(2\theta) = 2\cos^2 \theta - 1$, so let $\theta = x$.

453. $\dfrac{\cos(2x)}{\sin(2x)} = \dfrac{2\cos^2 x - 1}{2\sin x \cos x}$

$\qquad = \dfrac{2\cos^2 x - (\sin^2 x + \cos^2 x)}{2\sin x \cos x}$

$\qquad = \dfrac{\cos^2 x - \sin^2 x}{2\sin x \cos x}$

$\qquad = \dfrac{\dfrac{\cos^2 x - \sin^2 x}{\sin x \cos x}}{2}$

$\qquad = \dfrac{\dfrac{\cos x}{\sin x} - \dfrac{\sin x}{\cos x}}{2}$

$\qquad = \dfrac{\cot x - \tan x}{2}$

By questions 449 and 450.

454. $\tan(2\theta) = \tan(\theta + \theta)$

$\qquad = \dfrac{\tan\theta + \tan\theta}{1 - \tan\theta\tan\theta}$

$\qquad = \dfrac{2\tan\theta}{1 - \tan^2 \theta}$

455. $\tan(2x) = \dfrac{\sin(2x)}{\cos(2x)}$

$\qquad = \dfrac{2\sin x \cos x}{2\cos^2 x - 1}$

$\qquad = \dfrac{2\sin x \cos x}{2\cos^2 x - (\sin^2 x + \cos^2 x)}$

$\qquad = \dfrac{2\sin x \cos x}{\cos^2 x - \sin^2 x}$

By questions 449 and 450.

456. $\cot\left(\dfrac{x}{2}\right) = \dfrac{1}{\tan\left(\dfrac{x}{2}\right)}$

$= \dfrac{1}{\dfrac{1-\cos x}{\sin x}}$

$= \dfrac{\sin x}{1-\cos x}$

$= \dfrac{\sin x}{1-\cos x} \cdot \dfrac{1+\cos x}{1+\cos x}$

$= \dfrac{\sin x(1+\cos x)}{1-\cos^2 x}$

$= \dfrac{\sin x(1+\cos x)}{\sin^2 x}$

$= \dfrac{1+\cos x}{\sin x}$

Notice the strategy of multiplying by

$1\left(\dfrac{1+\cos x}{1+\cos x} = 1\right).$

457. $\dfrac{1-\tan^2 2x}{2\tan 2x} = \dfrac{1}{\dfrac{2\tan 2x}{1-\tan^2 2x}}$

$= \dfrac{1}{\tan(2(2x))}$

$= \dfrac{1}{\tan 4x}$

$= \cot 4x$

458. $\sin^2\left(\dfrac{x}{2}\right) = \dfrac{1-\cos x}{2}$

$= \dfrac{1-\dfrac{1}{\sec x}}{2}$

$= \dfrac{\sec x-1}{2\sec x}$

459. To verify that the statement is not an identity, show a counterexample. Let $\theta = 0°$. Then $\cos(2\theta) = \cos(2\cdot 0°) = \cos(0°) = 1$, but $2\cos\theta = 2\cos 0° = 2(1) = 2 \neq 1$. Thus, $\cos(2\theta) = 2\cos\theta$ is not an identity.

460. Let $\theta = 30°$ and $\beta = 60°$. Then $\sin(\theta + \beta) = \sin(30° + 60°) = \sin 90° = 1$, but $\sin\theta + \sin\beta = \sin(30°) + \sin(60°) = \dfrac{1}{2} + \dfrac{\sqrt{3}}{2} \neq 1$. Thus $\sin(\theta + \beta) = \sin\theta + \sin\beta$ is not an identity.

461. $2\sin\left(\dfrac{\pi}{3}\right) = 2\left(\dfrac{\sqrt{3}}{2}\right) = \sqrt{3}$. Yes, it is a solution.

462. $2\sec\left(\dfrac{\pi}{3}\right) = \dfrac{2}{\cos\left(\dfrac{\pi}{3}\right)} = \dfrac{2}{\left(\dfrac{1}{2}\right)} = 4$, but $\tan\left(\dfrac{\pi}{3}\right) + \cot\left(\dfrac{\pi}{3}\right) = \sqrt{3} + \dfrac{1}{\sqrt{3}} = \dfrac{4}{\sqrt{3}} \neq 4$.

Thus, it is not a solution.

463. Solve for $\sin x$,
$2\sin x - 1 = 0$

$\quad 2\sin x = 1$

$\quad \sin x = \dfrac{1}{2}$

Now, determine all values for x in $[0, 2\pi)$ for which $\sin x = \dfrac{1}{2}$. The reference angle is $\dfrac{\pi}{6}$.

The sine function is positive in quadrants I and II. Thus, $x = \dfrac{\pi}{6}, \dfrac{5\pi}{6}$.

464. $\sin x + \cos x = 0$

$\quad\quad\quad \sin x = -\cos x$

$\quad\quad\quad \dfrac{\sin x}{\cos x} = \dfrac{-\cos x}{\cos x} \quad\quad\quad\quad \text{provided } \cos x \neq 0.$

$\quad\quad\quad \tan x = -1$

Now, determine all values for x in $(-\infty, \infty)$ for which $\tan x = -1$. The reference angle is $\dfrac{\pi}{4}$.

The tangent function is negative in quadrants II and IV. Thus, $\dfrac{3\pi}{4}$ and $\dfrac{7\pi}{4}$ are the only two values in $[0, 2\pi)$ for which $\tan x = -1$. Because the tangent function has a period of π, all solutions to $\sin x + \cos x = 0$ can be expressed as $x = \dfrac{3\pi}{4} + n\pi$ and $x = \dfrac{7\pi}{4} + n\pi$, where n is an integer. *Note:* Because $\cos\dfrac{3\pi}{4} = -\dfrac{\sqrt{2}}{2} \neq 0$ and $\cos\dfrac{7\pi}{4} = \dfrac{\sqrt{2}}{2} \neq 0$, division by $\cos x$ in this problem was permissible.

465. $\sin x \tan x = \sin x$

$\sin x \tan x - \sin x = 0$

$\sin x(\tan x - 1) = 1$

$\sin x = 0$ or $\tan x - 1 = 0$

$\sin x = 0$ or $\tan x = 1$

Now, determine all values for x in $[0, 2\pi)$ for which either $\sin x = 1$ or $\tan x = 1$. For $\sin x = 0$, the reference angle is 0, so 0 and π are the only two values in $[0, 2\pi)$ for which $\sin x = 0$.

For $\tan x = 1$, the reference angle is $\dfrac{\pi}{4}$. The tangent function is positive in quadrants I and III, so $\dfrac{\pi}{4}$ and $\dfrac{5\pi}{4}$ are the only two values in $[0, 2\pi)$ for which $\tan x = 1$. Therefore, $\sin x \tan x = \sin x$ has solutions $x = 0, \dfrac{\pi}{4}, \pi, \dfrac{5\pi}{4}$.

466. $4\cos^2 x - 3 = 0$

$4\cos^2 x = 3$

$\cos^2 x = \dfrac{3}{4}$

$\cos x = \pm\dfrac{\sqrt{3}}{2}$

Now, determine all values for x in $[0, 2\pi)$ for which either $\cos x = \dfrac{\sqrt{3}}{2}$ or $\cos x = -\dfrac{\sqrt{3}}{2}$. The reference angle is $\dfrac{\pi}{6}$. The solutions are $x = \dfrac{\pi}{6}, \dfrac{5\pi}{6}, \dfrac{7\pi}{6}, \dfrac{11\pi}{6}$.

467. $\cos^3 x + \cos x = 0$

$\cos x(\cos^2 x + 1) = 1$

$\cos x = 0$ or $\cos^2 x + 1 = 0$

$\cos x = 0$ or $\cos^2 x = -1$ (no solution)

Now, determine all values for x in $[0, 2\pi)$ for which $\cos x = 0$. The reference angle is $\dfrac{\pi}{2}$, so the solutions are $x = \dfrac{\pi}{2}, \dfrac{3\pi}{2}$.

468.
$$\tan x + 3\cot x = 4$$
$$\tan x + \frac{3}{\tan x} = 4$$
$$\tan^2 x + 3 = 4\tan x$$
$$\tan^2 x - 4\tan x + 3 = 0$$
$$(\tan x - 3)(\tan x - 1) = 0$$
$$\tan x = 3 \text{ or } \tan x = 1$$

Now, determine all values for x in $[0°, 360°)$ for which either $\tan x = 3$ or $\tan x = 1$. For $\tan x = 3$, the reference angle is $x = 71.57°$. The tangent function is positive in quadrants I and III, so $71.57°$ and $251.57°$ are the only two values in $[0°, 360°)$ for which $\tan x = 3$. For $\tan x = 1$, the reference angle is $45°$. The tangent function is positive in quadrants I and III, so $45°$ and $225°$ are the only two values in $[0°, 360°)$ for which $\tan x = 1$. Therefore, $\tan x + 3\cot x = 4$ has solutions $x = 45°,\ 71.57°,\ 225°,\ 251.57°$.

469. $\cos x = 1 + \sqrt{3}\sin x$

$\cos x - 1 = \sqrt{3}\sin x$

Squaring both sides,

$(\cos^2 x - 1)^2 = (\sqrt{3}\sin x)^2$

$\cos^2 x - 2\cos x + 1 = 3\sin^2 x.$

Substituting $\sin^2 x = 1 - \cos^2 x,$

$\cos^2 x - 2\cos x + 1 = 3(1 - \cos^2 x)$

$\cos^2 x - 2\cos x + 1 = 3 - 3\cos^2 x$

$4\cos^2 x - 2\cos x - 2 = 0$

$2(2\cos x + 1)(\cos x - 1) = 0$

$2\cos x + 1 = 0 \text{ or } \cos x - 1 = 0$

$\cos x = -\dfrac{1}{2} \text{ or } \cos x = 1$

Caution: Squaring both sides can introduce extraneous roots.

Now, determine all values for x in $[0, 2\pi)$ for which either $\cos x = -\dfrac{1}{2}$ or $\cos x = 1$. For $\cos x = -\dfrac{1}{2}$, the reference angle is $\dfrac{\pi}{3}$, The cosine function is negative in quadrants II and III, so $\dfrac{2\pi}{3}$ and $\dfrac{4\pi}{3}$ are the only two values in $[0, 2\pi)$ for which $\cos x = -\dfrac{1}{2}$. For $\cos x = 1$, the reference angle is 0, so 0 is the only value in $[0, 2\pi)$ for which $\cos x = 1$. Therefore, tentatively $\cos x = 1 + \sqrt{3}\sin x$ has solutions $x = 0,\ \dfrac{2\pi}{3},\ \dfrac{4\pi}{3}$. The solutions are "tentative" because you need to make sure there are no extraneous roots, so check all solutions.

Checking $x = 0$: $\cos(0) = 1$ and $1 + \sqrt{3} \sin(0) = 1 + 0 = 1$, so 0 is a solution.

Checking $x = \dfrac{2\pi}{3}$: $\cos\left(\dfrac{2\pi}{3}\right) = -\dfrac{1}{2}$ and $1 + \sqrt{3} \sin\left(\dfrac{2\pi}{3}\right) = 1 + \sqrt{3}\left(\dfrac{\sqrt{3}}{2}\right) \neq -\dfrac{1}{2}$, so $\dfrac{2\pi}{3}$ is not a solution.

Checking $x = \dfrac{4\pi}{3}$: $\cos\left(\dfrac{4\pi}{3}\right) = -\dfrac{1}{2}$ and $1 + \sqrt{3} \sin\left(\dfrac{4\pi}{3}\right) = 1 + \sqrt{3}\left(-\dfrac{\sqrt{3}}{2}\right) = -\dfrac{1}{2}$, so $\dfrac{4\pi}{3}$ is a solution.

Thus, $\cos x = 1 + \sqrt{3} \sin x$ has solutions $x = 0$, $\dfrac{4\pi}{3}$.

---

470. Because the problem requires that $0 \le x < 2\pi$, it follows that $0 \le \dfrac{x}{2} < \pi$. Now, determine all values for $\dfrac{x}{2}$ between 0 and π for which $\cos\left(\dfrac{x}{2}\right) = \dfrac{1}{2}$. The reference angle is $\dfrac{\pi}{3}$. The cosine function is positive in quadrants I and IV. The solution then is $\dfrac{x}{2} = \dfrac{\pi}{3}$ or $x = \dfrac{2\pi}{3}$.

471. $\sin(2x) = \cos(2x)$

$\dfrac{\sin(2x)}{\cos(2x)} = 1$

$\tan(2x) = 1$

Because the problem requires that $0 \le x < 2\pi$, it follows that $0 \le 2x < 4\pi$. Now, determine all values for $2x$ between 0 and 4π for which $\tan(2x) = 1$. The reference angle is $\dfrac{\pi}{4}$. The tangent function is positive in quadrants I and III. The solutions then are $2x = \dfrac{\pi}{4}, \dfrac{5\pi}{4}, \dfrac{9\pi}{4}, \dfrac{13\pi}{4}$;

and, thus, $x = \dfrac{\pi}{8}, \dfrac{5\pi}{8}, \dfrac{9\pi}{8}, \dfrac{13\pi}{8}$.

472. $2 \cos x + \sin x = 1$

$2 \cos x = 1 - \sin x$

Squaring both sides,

$(2 \cos x)^2 = (1 - \sin x)^2$

$4 \cos^2 x = 1 - 2 \sin x + \sin^2 x$

Substituting $\cos^2 x = 1 - \sin^2 x$,

$4(1 - \sin^2 x) = 1 - 2 \sin x + \sin^2 x$

$4 - 4 \sin^2 x = 1 - 2 \sin x + \sin^2 x$

$5 \sin^2 x - 2 \sin x - 3 = 0$

$(5 \sin x + 3)(\sin x - 1) = 0$

$5 \sin x + 3 = 0$ or $\sin x = 1$

$\sin x = -\dfrac{3}{5}$ or $\sin x = 1$.

Solving $\sin x = -\dfrac{3}{5}$ yields $x = 216.87°$, $323.13°$. Solving $\sin x = 1$ yields $x = 90°$. Because squaring both sides might have introduced extraneous roots, check the solutions.

Checking $x = 90°$: $2\cos(90°) = 2(0) = 0$ and $1 - \sin(90°) = 1 - 1 = 0$, so $90°$ is a solution.

Checking $x = 216.87°$: $2\cos(216.87°) = -1.6$ and $1 - \sin(216.87°) = 1.6$, so $x = 216.87°$ is not a solution.

Checking $x = 323.13°$: $2\cos(323.13°) = 1.6$ and $1 - \sin(323.13°) = 1.6$, so $x = 323.13°$ is a solution.

Therefore, $2\cos x + \sin x = 1$ has solutions $x = 90°$, $323.13°$.

473. To solve a triangle means to determine the length of each side and the measure of each angle. Let $a =$ the length of the missing side, then using the Pythagorean theorem,

$a^2 = 14^2 - 10^2$

$a^2 = 196 - 100 = 96$

So, $a = \sqrt{96} \approx 9.8$.

$\cos A = \dfrac{\text{adjacent}}{\text{hypotenuse}} = \dfrac{10}{14} = 0.7143\ldots$ (don't round). Thus, using the $\cos^{-1}$ function, $\angle A \approx 44.4°$; and it follows that $\angle B \approx 90° - 44.4° = 45.6°$.

474. Let $a =$ the length of the missing side. This triangle is recognizable as a $30° - 60° - 90°$ right triangle, so $a = 1$, $\angle A = 30°$, and $\angle B = 60°$.

475. Let $a =$ the length of the missing side, then using the Pythagorean theorem,

$a^2 = 5^2 - 4^2$

$a^2 = 25 - 16 = 9$

So, $a = 3$.

$\sin A = \dfrac{\text{opposite}}{\text{hypotenuse}} = \dfrac{3}{5} = 0.6$. Thus, using the $\sin^{-1}$ function, $\angle A \approx 36.9°$, and hence $\angle B \approx 90° - 36.9° = 53.1°$.

476. Let $c =$ the length of the missing hypotenuse, then using the Pythagorean theorem,

$c^2 = 5^2 + 4^2$

$c^2 = 25 + 16 = 41$

So, $c = \sqrt{41} \approx 6.4$.

$\tan A = \dfrac{\text{opposite}}{\text{adjacent}} = \dfrac{4}{5} = 0.8$. Thus, using the $\tan^{-1}$ function, $\angle A \approx 38.7°$ and, hence, $\angle B \approx 90° - 38.7° = 51.3°$.

477. This "triangle" has no solution since the hypotenuse of a right triangle must be the longest side.

478. Since the sum of the angles of a triangle is $180°$, $\angle C = 180° - 38° - 64° = 78°$. Using the law of sines, $\dfrac{b}{\sin 64°} = \dfrac{75}{\sin 38°}$, so $b = \dfrac{75 \sin 64°}{\sin 38°} \approx 109.5$. Again, applying the law of sines, $\dfrac{c}{\sin 78°} = \dfrac{75}{\sin 38°}$, so $c = \dfrac{75 \sin 78°}{\sin 38°} \approx 119.2$.

479. $\angle C = 180° - 13° - 22° = 145°$. Using the law of sines, $\dfrac{a}{\sin 13°} = \dfrac{126}{\sin 145°}$, so $a = \dfrac{126 \sin 13°}{\sin 145°} \approx 49.4$. Again, applying the law of sines $\dfrac{b}{\sin 22°} = \dfrac{126}{\sin 145°}$, so $b = \dfrac{126 \sin 22°}{\sin 145°} \approx 82.3$.

480. $\angle A = 180° - 103.4° - 19.6° = 57°$. Using the law of sines $\dfrac{b}{\sin 103.4°} = \dfrac{42.7}{\sin 57°}$, so $b = \dfrac{42.7 \sin 103.4°}{\sin 57°} \approx 49.5$. Again, applying the law of sines $\dfrac{c}{\sin 19.6°} = \dfrac{42.7}{\sin 57°}$, so $c = \dfrac{42.7 \sin 19.6°}{\sin 57°} \approx 17.1$.

481. Using the law of cosines, $a^2 = (3.2)^2 + (1.5)^2 - 2(3.2)(1.5) \cos 95.7° = 13.4434\ldots$. Thus, $a = \sqrt{13.4434\ldots} = 3.6665\ldots \approx 3.7$. Now, using the law of sines, $\sin B = \dfrac{b \sin A}{a} = \dfrac{3.2 \sin 95.7°}{3.6665\ldots} \approx 0.8684\ldots$. Hence, $\angle B \approx 60.3°$ and $\angle C \approx 180° - 95.7° - 60.3° = 24.0°$.

482. Using the law of cosines, $c^2 = 75^2 + 32^2 - 2(75)(32) \cos 38° \approx 2866.5483\ldots$. Thus, $c = \sqrt{2866.5483\ldots} = 53.5401\ldots \approx 53.5$. Now using the law of sines, $\sin B = \dfrac{32 \sin 38°}{53.5401\ldots} \approx 0.3679\ldots$. Hence, $\angle B \approx 21.6°$ and $\angle A \approx 180° - 21.6° - 38° = 120.4°$.

Chapter 11: Conic Sections

483. $x^2 + y^2 = 1$

The circle is the unit circle with center $(0, 0)$ and radius $=1$. See the figure below.

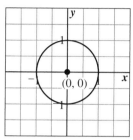

484. $x^2 - 4x + 2y + y^2 = 4$

Write the equation in standard form:

$(x^2 - 4x + \quad) + (y^2 + 2y + \quad) = 4$

Complete the squares inside the parentheses and balance the equation:

$(x^2 - 4x + 4) + (y^2 + 2y + 1) = 4 + 4 + 1$

Factor and collect terms:

$(x - 2)^2 + (y + 1)^2 = 9$

Thus, the circle has center $(2, -1)$ with radius $= \sqrt{9} = 3$. See the figure below.

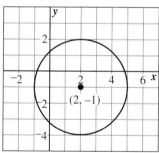

485. Proceed as in 484.

$$x^2 + y^2 - 4x + 10y = -4$$
$$(x^2 - 4x + \quad) + (y^2 + 10y + \quad) = -4$$
$$(x^2 - 4x + 4) + (y^2 + 10y + 25) = -4 + 4 + 25$$
$$(x - 2)^2 + (y + 5)^2 = 25$$

Thus, the circle has center $(2, -5)$ with radius $= \sqrt{25} = 5$.

486. Proceed as in 484.

$$3x^2 + 6x + 3y^2 - 12y = 49$$

$$3(x^2 + 2x + \quad) + 3(y^2 - 4y + \quad) = 49$$

$$3(x^2 + 2x + 1) + 3(y^2 - 4y + 4) = 49 + 3 + 12$$

$$3(x^2 + 2x + 1) + 3(y^2 - 4y + 4) = 64$$

$$3(x + 1)^2 + 3(y - 2)^2 = 64$$

$$(x + 1)^2 + (y - 2)^2 = \frac{64}{3}$$ *Note:* Divide by 3 to put the equation in standard form.

Thus, the circle has center at $(-1, 2)$ and radius $= \sqrt{\dfrac{64}{3}} = \dfrac{8}{\sqrt{3}} = \dfrac{8\sqrt{3}}{3}$.

487. $(x + 4)^2 + (y + 7)^2 = (\sqrt{11})^2$ or $(x + 4)^2 + (y + 7)^2 = 11$

488. Substitute the coordinates successively into the standard form $(x - h)^2 + (y - k)^2 = r^2$:

Using $(0, 0)$ and simplifying yields

$(0 - h)^2 + (0 - k)^2 = r^2$

Equation 1: $h^2 + k^2 = r^2$

Using $(0, 2)$ and simplifying yields

$(0 - h)^2 + (2 - k)^2 = r^2$

Equation 2: $h^2 + 4 - 4k + k^2 = r^2$

Using $(2, 0)$ yields

$(2 - h)^2 + (0 - k)^2 = r^2$

Equation 3: $4 - 4h + h^2 + k^2 = r^2$

Because the right sides of Eqs. 1 and 2 both equal r^2, set the left sides equal to each other and simplify:

$h^2 + k^2 = h^2 + 4 - 4k + k^2$

$4k = 4$

Solve for k:

$k = 1$

Similarly, because the right sides of Eqs. 1 and 3 both equal r^2, set the left sides equal to each other and simplify:

$h^2 + k^2 = 4 - 4h + h^2 + k^2$

$4h = 4$

Solve for h:

$h = 1$

Now, solve for r^2 by substituting $h = 1$ and $k = 1$ into Eq. 1:

$$1^2 + 1^2 = r^2$$
$$2 = r^2$$

Thus, the circle has equation $(x - 1)^2 + (y - 1)^2 = 2$.

489. $4x^2 + 25y^2 = 100$

Divide by 100 to put the equation in standard form:

$$\frac{x^2}{25} + \frac{y^2}{4} = 1$$

Thus, the ellipse has center $(0, 0)$ and vertices $(h \pm a, k) = (0 \pm \sqrt{25}, 0) = (\pm 5, 0)$ with horizontal major axis of length $= 2a = 2\sqrt{25} = 2 \cdot 5 = 10$ and minor axis of length $= 2b = 2\sqrt{4} = 2 \cdot 2 = 4$. See the figure below.

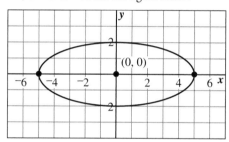

490. $9x^2 + 4y^2 = 36$

Divide by 36 to put the equation in standard form:

$$\frac{x^2}{4} + \frac{y^2}{9} = 1$$

Thus, the ellipse has center $(0, 0)$ and vertices $(h, k \pm a) = (0, 0 \pm \sqrt{9}) = (0, \pm 3) = (0, 3)$ and $(0, -3)$ with vertical major axis of length $= 2a = 2\sqrt{9} = 2 \cdot 3 = 6$ and minor axis of length $= 2b = 2\sqrt{4} = 2 \cdot 2 = 4$. See the figure below.

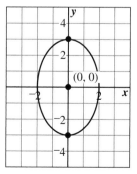

491. $x^2 + 4y^2 - 8y + 4x - 8 = 0$

Write the equation in standard form:
$$x^2 + 4x + 4y^2 - 8y = 8$$
$$(x^2 + 4x + \quad) + 4(y^2 - 2y + \quad) = 8$$
$$(x^2 + 4x + 4) + 4(y^2 - 2y + 1) = 8 + 4 + 4$$
$$(x + 2)^2 + 4(y - 1)^2 = 16$$
$$\frac{(x + 2)^2}{16} + \frac{(y - 1)^2}{4} = 1$$

Thus, the ellipse has center $(-2, 1)$ and vertices $(h \pm a, k) = (-2 \pm \sqrt{16}, 1) = (-2 \pm 4, 1) = (-6, 1)$ and $(2, 1)$ with horizontal major axis of length $= 2a = 2\sqrt{16} = 2 \cdot 4 = 8$ and minor axis of length $= 2b = 2\sqrt{4} = 2 \cdot 2 = 4$.

492. $5x^2 + 2y^2 + 20y - 30x + 75 = 0$

Write the equation in standard form:
$$5x^2 - 30x + 2y^2 + 20y = -75$$
$$5(x^2 - 6x + \quad) + 2(y^2 + 10y + \quad) = -75$$
$$5(x^2 - 6x + 9) + 2(y^2 + 10y + 25) = -75 + 45 + 50$$
$$5(x - 3)^2 + 2(y + 5)^2 = 20$$
$$\frac{(x - 3)^2}{4} + \frac{(y + 5)^2}{10} = 1$$

Thus, the ellipse has center $(3, -5)$ and vertices $(h, k \pm a) = (3, -5 \pm \sqrt{10}) = (3, -5 + \sqrt{10})$ and $(3, -5 - \sqrt{10})$ with vertical major axis of length $= 2a = 2\sqrt{10}$ and minor axis of length $= 2b = 2\sqrt{4} = 2 \cdot 2 = 4$.

493. $2x^2 + 5y^2 - 12x + 20y - 12 = 0$

Write the equation in standard form:
$$2x^2 - 12x + 5y^2 + 20y = 12$$
$$2(x^2 - 6x + \quad) + 5(y^2 + 4y + \quad) = 12$$
$$2(x^2 - 6x + 9) + 5(y^2 + 4y + 4) = 12 + 18 + 20$$
$$2(x - 3)^2 + 5(y + 2)^2 = 50$$
$$\frac{(x - 3)^2}{25} + \frac{(y + 2)^2}{10} = 1$$

Thus, the ellipse has center $(3,-2)$ and vertices $(h \pm a, k) = (3 \pm \sqrt{25}, -2) = (3 \pm 5, -2) = (8,-2)$ and $(-2,-2)$ with horizontal major axis of length $= 2a = 2\sqrt{25} = 2 \cdot 5 = 10$ and minor axis of length $= 2b = 2\sqrt{10}$.

494. $36x^2 - 16y^2 = 144$

Divide by 144 to put the equation in standard form:

$$\frac{x^2}{4} - \frac{y^2}{9} = 1$$

Thus, the hyperbola has center $(0,0)$, vertices $(h \pm a, k) = (0 \pm \sqrt{4}, 0) = (\pm 2, 0) = (-2, 0)$ and $(2,0)$ and asymptotes $y = \pm \frac{b}{a}(x - h) + k = \pm \frac{\sqrt{9}}{\sqrt{4}}(x - 0) - 0 = \pm \frac{3}{2}x$ with horizontal transverse axis of length $= 2a = 2\sqrt{4} = 2 \cdot 2 = 4$ and conjugate axis of length $= 2b = 2\sqrt{9} = 2 \cdot 3 = 6$. See the figure below.

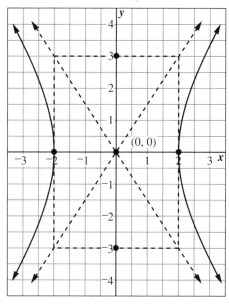

495. $16y^2 - 4x^2 + 24x = 100$

Write the equation in standard form:

$$16y^2 - 4(x^2 - 6x + \quad) = 100$$
$$16y^2 - 4(x^2 - 6x + 9) = 100 - 36$$
$$16y^2 - 4(x - 3)^2 = 64$$
$$\frac{y^2}{4} - \frac{(x - 3)^2}{16} = 1$$

Thus, the hyperbola has center $(3,0)$, vertices $(h, k \pm a) = (3, 0 \pm \sqrt{4}) = (3, \pm 2) = (3, 2)$ and $(3, -2)$, and asymptotes $y = \pm \dfrac{a}{b}(x - 3) + k = \pm \dfrac{\sqrt{4}}{\sqrt{16}}(x - 3) + 0 = \pm \dfrac{1}{2}(x - 3)$ with vertical transverse axis of length $= 2a = 2\sqrt{4} = 2 \cdot 2 = 4$ and conjugate axis of length $= 2b = 2\sqrt{16} = 2 \cdot 4 = 8$. See the figure below.

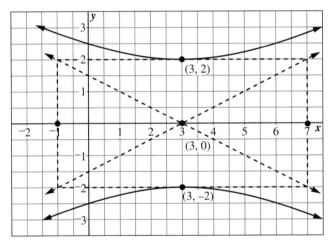

496. $9y^2 - x^2 + 54y + 4x + 68 = 0$

Write the equation in standard form:

$$9y^2 - x^2 + 54y + 4x = -68$$
$$9(y^2 + 6y + \quad) - 1(x^2 - 4x + \quad) = -68$$
$$9(y^2 + 6y + 9) - 1(x^2 - 4x + 4) = -68 + 81 - 4$$
$$9(y + 3)^2 - 1(x - 2)^2 = 9$$
$$\frac{(y + 3)^2}{1} - \frac{(x - 2)^2}{9} = 1$$

Thus, the hyperbola has center $(2, -3)$, vertices $(h, k \pm a) = (2, -3 \pm \sqrt{1}) = (2, -3 \pm 1) = (2, -2)$ and $(2, -4)$, and asymptotes $y = \pm \frac{a}{b}(x - h) + k = \pm \frac{\sqrt{1}}{\sqrt{9}}(x - 2) - 3 = \pm \frac{1}{3}(x - 2) - 3$ with vertical transverse axis of length $= 2a = 2\sqrt{1} = 2 \cdot 1 = 2$ and conjugate axis of length $= 2b = 2\sqrt{9} = 2 \cdot 3 = 6$.

497. $4x^2 - 9y^2 - 24x + 72y - 144 = 0$

Write the equation in standard form:

$$4x^2 - 24x - 9y^2 + 72y = 144$$
$$4(x^2 - 6x + \quad) - 9(y^2 - 8y + \quad) = 144$$
$$4(x^2 - 6x + 9) - 9(y^2 - 8y + 16) = 144 + 36 - 144$$
$$4(x - 3)^2 - 9(y - 4)^2 = 36$$
$$\frac{(x - 3)^2}{9} - \frac{(y - 4)^2}{4} = 1$$

Thus, the hyperbola has center $(3, 4)$, vertices $(h \pm a, k) = (3 \pm \sqrt{9}, 4) = (3 \pm 3, 4) = (6, 4)$ and $(0, 4)$, and asymptotes $y = \pm \frac{b}{a}(x - h) + k = \pm \frac{\sqrt{4}}{\sqrt{9}}(x - 3) + 4 = \pm \frac{2}{3}(x - 3) + 4$ with horizontal transverse axis of length $= 2a = 2\sqrt{9} = 2 \cdot 3 = 6$ and conjugate axis of length $= 2b = 2\sqrt{4} = 2 \cdot 2 = 4$.

498. $x^2 = -12y$

Write the equation in standard form:

$(x - 0)^2 = -12(y - 0)$

Thus, the parabola has a vertical axis, and its vertex is (0, 0). Furthermore, $4p = -12$, giving $p = -3$. Hence, the focus is $(0, 0 - 3) = (0, -3)$ and the directrix is $y = 0 - (-3) = 3$. Because $p = -3 < 0$, the parabola opens downward. See the figure below.

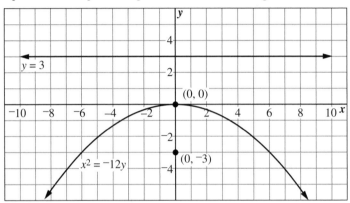

499. $y^2 = 4x$

Write the equation in standard form:

$(y - 0)^2 = 4(x - 0)$

Thus, the parabola has a horizontal axis, and its vertex is (0, 0). Furthermore, $4p = 4$, giving $p = 1$. Hence, the focus is $(0 + 1, 0) = (1, 0)$ and its directrix is $x = 0 - 1 = -1$. Because $p = 1 > 0$, the parabola opens to the right. See the figure below.

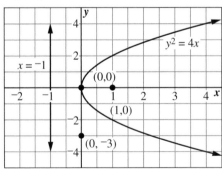

500. $y^2 + 2y + 16x = -49$

Write the equation in standard form:

$$(y^2 + 2y +) = -16x - 49$$
$$(y^2 + 2y + 1) = -16x - 49 + 1$$
$$(y + 1)^2 = -16x - 48$$
$$(y + 1)^2 = -16(x + 3)$$

Thus, the parabola has a horizontal axis, and its vertex is $(-3, -1)$. Furthermore, $4p = -16$, giving $p = -4$. Hence, the focus is $(-3 - 4, -1) = (-7, -1)$ and its directrix is $x = -3 - (-4) = 1$. Because $p = -4 < 0$, the parabola opens to the left.

501. $x^2 - 8y = 6x - 25$

Write the equation in standard form:

$$(x^2 - 6x +) = 8y - 25$$
$$(x^2 - 6x + 9) = 8y - 25 + 9$$
$$(x - 3)^2 = 8y - 16$$
$$(x - 3)^2 = 8(y - 2)$$

Thus, the parabola has a vertical axis, and its vertex is $(3, 2)$. Furthermore, $4p = 8$, giving $p = 2$. Hence, the focus is $(3, 2 + 2) = (3, 4)$ and the directrix is $y = 2 - 2 = 0$. Because $p = 2 > 0$, the parabola opens upward.